Global weather patterns revealed by picture from geostationary satellite above the equator over the eastern Pacific Ocean. (Courtesy, NASA)

THE ATMOSPHERIC SCIENCES: Problems and Applications

Committee on Atmospheric Sciences
Assembly of Mathematical and Physical Sciences
National Research Council

NATIONAL ACADEMY OF SCIENCES
Washington, D.C. 1977

NOTICE: The project that is the subject of this report was approved by the Governing Board of the National Research Council, whose members are drawn from the Councils of the National Academy of Sciences, the National Academy of Engineering, and the Institute of Medicine. The members of the Committee responsible for the report were chosen for their special competences and with regard for appropriate balance.

This report has been reviewed by a group other than the authors according to procedures approved by a Report Review Committee consisting of members of the National Academy of Sciences, the National Academy of Engineering, and the Institute of Medicine.

The activities of the Committee on Atmospheric Sciences leading to this report have been supported by the National Oceanic and Atmospheric Administration and the National Science Foundation under Contract NSF-C310, Task Order No. 9.

Library of Congress Cataloging in Publication Data

National Research Council. Committee on Atmospheric Sciences.
The atmospheric sciences.

Bibliography: p.
1. Weather. 2. Climatology. 3. Environmental policy. I. Title.
QC981.N37 1978 333.9 77-24640
ISBN 0-309-02626-1

Available from:

Printing and Publishing Office, National Academy of Sciences
2101 Constitution Avenue, N.W., Washington, D.C. 20418

Printed in the United States of America

Committee on Atmospheric Sciences

GEORGE S. BENTON, The Johns Hopkins University, *Chairman*
VERNER E. SUOMI, University of Wisconsin, *Vice Chairman*
DAVID ATLAS, National Center for Atmospheric Research
LOUIS J. BATTAN, University of Arizona
FRANCIS P. BRETHERTON, University Corporation for Atmospheric Research
HARVEY BROOKS, Harvard University
THOMAS M. DONAHUE, University of Michigan
W. LAWRENCE GATES, Oregon State University
RICHARD M. GOODY, Harvard University
CHARLES L. HOSLER, Pennsylvania State University
ROBERT W. KATES, Clark University
J. MURRAY MITCHELL, JR., National Oceanic and Atmospheric Administration
JOSEPH PEDLOSKY, University of Chicago
NORMAN J. ROSENBERG, University of Nebraska
JOHN M. WALLACE, University of Washington
PAUL E. KLOPSTEG, Glenview, Illinois, *member emeritus*
THOMAS F. MALONE, Butler University, *ex officio*

JOHN R. SIEVERS, National Research Council, *Executive Secretary*

Preface

Over the years, the Committee on Atmospheric Sciences of the National Research Council has published scientific and technical reports dealing with many aspects of the atmospheric sciences. The reports have dealt primarily with specific areas of research and application and have been addressed to an audience of scientists and administrators familiar with the subjects. The procedure commonly followed has involved the establishment of a panel of experts, who would review a particular activity and formulate sets of findings, conclusions, and recommendations intended to serve as guides for the scientific community and advice to the federal government. The resulting documents usually were directed primarily to those federal agencies that support research in the particular areas of interest of each report. As examples, we can cite publications dealing with the following subjects: atmospheric chemistry, weather and climate modification, remote probing of the atmosphere, education and manpower, severe storms, global observations and numerical simulation, and climate.

The Committee on Atmospheric Sciences has sought to extend the use of its reports beyond the several executive departments and agencies conducting research in the atmospheric sciences. Copies are distributed to private and academic research centers and to congressional officials as well.

It has been apparent to the Committee that the technical nature of its reports has tended to restrict the reading audience. Many individuals in government and universities and among the informed citizenry are not familiar with the intellectual challenges of the atmospheric sciences and the crucial effects of the atmosphere on most human affairs. Occasional magazine and newspaper articles deal with one or more aspects of the weather and climate. Some such stories are well written and factually correct; others can only mislead the reader. The scope and progress of the atmospheric sciences and the value of greater knowledge to society as a whole have not been adequately covered in any publications currently available.

This report represents a modest first step in filling the information gap just described. The Committee on Atmospheric Sciences has attempted to summarize the types of problems that are encountered in atmospheric research and to show why such problems are indeed scientific problems of an involved and complex nature, yet, for the most part, solvable, providing sufficient attention and manpower can be directed toward them.

Although the study of the atmosphere is one of the oldest of sciences, having been practiced by Aristotle, who in about 340 B.C. wrote *Meteorologica*, giving his views on weather and climate, the major advances have occurred during this century, particularly over the last few decades.

Significant progress in understanding the atmosphere has been made in almost steplike fashion following the development of new techniques for observing the atmosphere and gathering and analyzing the data. The introduction of the telegraph, in the middle of the last century, made it possible to assemble observations from distant places and construct weather maps rapidly. This development led to improved, timely descriptions of the atmosphere and better weather forecasts.

The invention in 1927 of the radiosonde, an effective balloonborne instrument for measuring atmospheric pressure, temperature, and humidity, has led to greatly improved descriptions of the three-dimensional structure of the atmosphere and to a greater knowledge of fronts, storm systems, jet streams, and many other atmospheric phenomena.

Radar developments during World War II allowed measurements of important properties of thunderstorms, hurricanes, and other precipitating storms. Rockets, whose scientific potential was demonstrated following World War II, have been used to probe the upper atmosphere and learn about the ozonosphere and ionosphere.

METEOROLOGICAL SATELLITE

The development of instrumented earth-orbiting satellites has been a continuing story of great importance to the atmospheric sciences. It is commonplace now to view the cloud cover over the entire planet and watch the growth and motion of hurricanes and other weather disturbances. By means of radiometric techniques, it is possible to estimate temperatures and other properties of the air, temperatures of the ocean, ice cover over the earth, and other items of great geophysical interest.

Satellites also are playing vital roles in the transmission of observations from many hundreds of weather stations and other observing platforms to weather centers in various countries.

COMPUTERS

Electronic computers, a product of scientific and engineering genius of the last three decades, have revolutionized the atmospheric sciences as they have other sciences as well. When employed by skilled scientists, computers and associated peripheral equipment can, in reasonable times, assimilate data from stations all over the globe, analyze them, and depict patterns of wind, pressure, and other phenomena. By means of computers, atmospheric scientists have been able to deal, by numerical means, with the equations of motion of the atmosphere over the earth. Computers, more than any other tool, have transformed weather forecasting from an art to a science.

With the improvements in observational and computational technology, there has been a steady growth in the number of competent atmospheric scientists. They have taken on research programs of increasing scales of difficulty and complexity.

RESEARCH ON VIOLENT STORMS

One of the first large meteorological research programs, called the Thunderstorm Project, was carried out in the late 1940's under the direction of Horace R. Byers. A very ambitious program by standards of the day, the Thunderstorm Project involved many people, airplanes, and radars and a host of other instruments. The project removed some of the mystery of thunderstorms by flying through them and by probing them from below. Much more has subsequently been learned about thunderstorms, but many features of the dynamics and thermodynamics of such storms still remain to be better understood.

Public interest in severe thunderstorms, particularly those producing hail and tornadoes, has always been high. Violent weather does widespread damage to agriculture and property and, more importantly, causes injuries and deaths. Hurricanes periodically visit populated areas of the world and occasionally cause enormous losses of lives and properties. Over the last few decades or so, there has been considerable research on the nature and prediction of violent weather. Since the early 1970's, the National Center for Atmospheric Research (NCAR), with the collaboration of many university scientists, has been the focus of research on hailstorms, and NCAR is currently working with the National Oceanic and Atmospheric Administration (NOAA) in the design of a project (called SESAME) to investigate weather systems that produce severe storms over the Great Plains of the United States.

CLOUD PHYSICS AND WEATHER MODIFICATION

In the late 1940's and early 1950's, there were major advances in understanding of the physics of clouds and precipitation and the use of cloud-seeding techniques for modifying them. Irving Langmuir and Vincent J. Schaefer at General Electric Research Laboratories were pioneers in these endeavors. Over the intervening years, there has been significant progress and some disappointments on the road to the development of a reliable technology for modifying the weather.

THE UPPER ATMOSPHERE

The International Geophysical Year (IGY), involving worldwide efforts devoted to the study of the earth, was a highlight of geophysical activities in the late 1950's. Research started at that time has taught us many things about the ozone layer and the properties and processes of the upper atmosphere. The more recent International Magnetospheric Study is substantially increasing our knowledge of the very highest reaches of the earth's atmosphere dominated by energetic, charged particles and the magnetic fields of the earth. The IGY also gave special attention to polar regions of the earth and the roles they play in influencing weather and climate.

AIR QUALITY

The 1960's saw a greater awareness of the hazards of environmental pollution, particularly of water and air. This has led to improved techniques for monitoring atmospheric contaminants. Networks of stations now make systematic observations of the chemical constituents of air and of rainwater.

Over the last few years, the central concern of a number of atmospheric chemists has shifted to the upper atmosphere. Possible effects on the ozone layer, caused by engine emissions from high-flying airplanes and of chlorofluoromethane propellants from spray cans have been getting serious attention. It still is not clear to what extent the quantity of ozone between about 20 and 50 kilometers above the ground is likely to be diminished by the introduction of man-made substances. There still is much to learn about the relationship between atmospheric ozone and skin cancer and other biological phenomena and whether chlorofluoromethanes in the atmosphere can affect the climate. Much is at stake, and answers need to be found fairly quickly.

SEASONAL WEATHER AND CLIMATE

The number of people on the earth continues to increase at a rate that threatens to outstrip the capacity of the food production systems of this planet. Over the last decade or so, widespread droughts, particularly in Africa and the Soviet Union, have demonstrated the degree to which unfavorable weather can reduce food production and cause human misery.

Scientists have been aware that, on a worldwide basis, average air temperatures have undergone small but significant changes over periods of several decades. It also has been known that there has been, for more than 80 years, a gradual increase in atmospheric carbon dioxide, caused largely by the combustion of fossil fuels. It has been argued that the increases in carbon dioxide and possible increases in smoke and dust particles are affecting the global climate. It even has been speculated that unusual seasonal weather in India, Africa, and elsewhere might be explained in terms of man-induced climate changes.

There is no doubt that seasonal weather is normally variable. Droughts and floods have always been with us and will be in the future as well. Similarly, the climate of a region or of the entire earth undergoes fluctuations over periods of decades as well as much longer periods. An important task confronting atmospheric scientists is that of learning the reasons for these changes. Are they chance variations whose occurrence can only be predicted in probabilistic terms? Or can they be explained in a deterministic way, in terms of some external (perhaps from the sun) or internal (such as environmental pollution) forcing functions?

Supplies of food and freshwater depend in a crucial way on weather and climate. Energy consumption does likewise. The dependence will increase as global population increases. Greater understanding of seasonal weather and climate is essential.

WEATHER FORECASTING AND CLIMATE RESEARCH

The 1960's saw the inception of the World Weather Programs. They included the World Weather Watch, a program which the weather services of the world, under the auspices of the World Meteorological Organization, seek to improve the quality of weather forecasts and other services to all countries.

The research component of the World Weather Programs is the Global Atmospheric Research Program (GARP), an international effort aimed at extending the period and the accuracy of weather forecasts and developing an improved understanding of global climate. Subprograms of GARP have included extensive research endeavors in the area of Barbados (in 1969) and across the tropical ocean between South America and Africa (in 1974). These projects have sought to learn about interactions of the atmosphere and the oceans and the role of the tropics in the maintenance of global air circulation patterns. As part of the buildup of knowledge to be used in pursuing the objectives of GARP, there are plans to conduct special research in polar regions and the Asian subcontinents.

In the latter part of the 1970's, the first global experiment of the GARP will be conducted. It will involve a large array of modern equipment, supplied and manned by scientists and engineers from many countries. Increases in observations in time and space are only the starting point toward a better understanding of the atmosphere and its coupling with the oceans covering the Planet Earth. This goal can only be achieved through the efforts of competent, dedicated scientists.

CONCLUSION

We and many generations to follow will have only Planet Earth to live on. Research programs that will teach us the secrets of its land, oceans, and atmosphere will serve us well. Obvious tangible benefits of greater knowledge of the atmosphere should be better forecasts and many other weather information services. Other benefits should accrue from the development of a body of information that can be used by individuals engaged in agriculture; in the supply of water to cities, farms, and industry; in the extraction of resources from the seas; in oceanic and air transport; and in the maintenance of a healthy environment for all the earth's inhabitants.

ACKNOWLEDGMENT

A large number of individuals contributed to the substance and organization of this publication. There are some to whom particular thanks are due for their early and initial interest and assistance in the development of the various sections and topics of the report. On behalf of the Committee on Atmospheric Sciences it is a special privilege to note the major contributions that have been made by the following scientists: Stanley A. Changnon, Jr., Illinois State Water Survey; Thomas M. Donahue, University of Michigan; Robert G. Fleagle, University of Washington; W. Lawrence Gates, Oregon State University; Edwin Kessler, National Severe Storms Laboratory (NOAA); Thomas F. Malone, Butler University; Donald H. Pack, formerly with the Air Resources Laboratory (NOAA); John S. Perry, National Research Council; Norman J. Rosenberg, University of Nebraska; and Jack C. Thompson, San Jose State University.

Louis J. Battan
Committee on Atmospheric Sciences

Contents

1

The Atmospheric Sciences and Society

In addressing the interrelationship of the atmospheric sciences and society, it is appropriate to reflect briefly on the predicaments of the contemporary state of affairs, to identify a few of the major issues, and then to examine the relevance of the atmospheric sciences to these issues.

In these days, when science and technology are so often viewed by the general public either as an esoteric occupation for a societal elite or as a cornucopia that can be turned on at short notice to spew forth instant solutions to the problems that confront society, it is important that we take a long-term perspective. Let us examine, briefly, a few aspects of the earth's geophysical history.

EVOLUTION OF THE ATMOSPHERE

According to the best scientific evidence and interpretation, Spaceship Earth was launched (in a manner not fully understood) nearly 5 billion years ago. The primordial atmosphere was devoid of free

---

Prepared by Thomas F. Malone, Butler University.

oxygen and consisted of a mixture of hydrogen, methane, ammonia, and water vapor, from which the amino acids and simpler organic compounds are synthesized by an energy source such as lightning or solar ultraviolet radiation. Volcanic eruptions added more water vapor, nitrogen, and carbon dioxide. Further cooling created the warm shallow seas. The processes of life were ignited more than three billion years ago in this warm oceanic broth in the form of anaerobic organisms that were dependent on ambient organic molecules and amino acids for nourishment and livelihood through simple fermentation. A giant step had been taken in the evolution of life on Planet Earth.

Within less than a billion years an equally significant step occurred. A few of the organisms succeeded in changing their mode of existence from one of dependence on fermentation to the more efficient mode of photosynthesis and respiration. This set the stage for the release of oxygen, the fixation of nitrogen, and further evolution, as cells developed the capacity to divide by mitosis—a necessary precondition for higher life. As organisms that could not tolerate free oxygen were replaced by more efficient respiring forms, oxygen-mediating enzymes appeared as the chemical catalysts governing metabolism and producing the structures prescribed by the DNA molecules.

The impact of these biological processes on the atmosphere was profound. They caused the oxygen content to increase and, under the impact of ultraviolet radiation, some oxygen molecules dissociated into highly reactive oxygen atoms and, particularly at upper levels, equally reactive ozone, which subsequently was to provide a protective shield favoring life forms with a low tolerance for ultraviolet radiation. The evidence is persuasive that early forms of life played an important role in determining the composition and the structure of our atmosphere.

Invertebrates appeared about 500 million years ago, vertebrates 400 million years ago. The dinosaurs reigned during the interval from 200 million years to 60 million years in the past. Man emerged less than 5 million years ago, and so-called Modern Man dates back several tens of thousands of years.

Against a summary of several billions of years of evolutionary history in a few sentences, what can be said of the future? It is clear that because of the remaining solar energy and the low probability of cosmic collisions, the earth *should* be habitable for at least another hundred million years. Can the human species, which is astute enough to use modern plate tectonics to predict the shape of the continents 25 million years from now, be sufficiently wise to survive that long? This issue may be resolved to a large extent—one way or another—during the next few decades.

CULTURAL REVOLUTIONS

In highly abbreviated form, four major developments—distinct transitions, if you will—of the past few thousand years provide some guidance for expectations of the future.

First came the rapid materialization of interest all over the world in a spiritual life, in a better order, in a reinterpretation of the relation between a Supreme External Force and men—and among men—that took place over a few millenia starting some 5000 years ago. Hunting had given way to agrarian practices; village-size communities had given way to urban-based kingdoms. Man had time to reflect on the reason for being, to raise the most fundamental of ethical questions, and to propound doctrinal responses. In one manner or another, great philosophers addressed in their own way the three questions summarized with such elegant simplicity by Immanuel Kant in his famous *Critique of Pure Reason*,* when he wrote that, ". . . the whole interest of reason, speculative as well as practical, is centered in the following questions: What *can* I know? What *ought* I to do? What *may* I hope?"

The second transition was the Copernican revolution in the sixteenth century, which revealed to us *where* we are—inhabitants on a small planet, circling about one of a great many stars in a vast universe.

The third major development was the Darwinian revolution, which told us *what* we are—complex products of a process that proceeded over several billion years from polyatomic molecules, to cells, to organs, to organisms, to colonies, packs, flocks, tribes, and nations.

The fourth development has been the scientific–technological–industrial revolution. By tapping the solar energy stored in fossil fuels and the power of the atom; by our ability to manipulate natural materials and to synthesize new materials; by an enhanced understanding of life processes, genetic laws, and the chemical nature of a gene; and by our recently acquired and rapidly expanding capacity to handle information, we are introducing significant perturbations in the relationship of humans and their natural environment. The implications of these perturbations are many:

- The capacity to use energy multiplies the work-performing capability of an individual hundreds of times, making possible modern transportation, construction, and manufacturing, for example. It also

*I. Kant (1902). *Critique of Pure Reason*. American Tome Library Co., New York, 617 pp.

threatens the world with nuclear destruction, dissipation of the ultraviolet-shielding layer of ozone in the upper atmosphere. A century or two of unrestrained growth in energy production could have a profound influence on the global climate.

• The capacity to manipulate materials provides a host of new and useful consumer goods, ranging from petrochemical products to hand computers. It supports the "Green Revolution" by providing new insecticides and fertilizers. It is also the source of air, water, and land pollution that pose serious threats to human life.

• The capacity to influence biological processes has led to new and more productive strains of grain and breeds of animals; it has prolonged human life. It has also aggravated the explosively expanding demands of more and more people for limited resources. It is posing problems of genetic engineering DNA recombination, cloning, and the ethical practice of medicine.

• The capacity to handle information may turn out to be the most portentous of all. It is possible to observe parts of the Universe veiled from the human eye, to manipulate machines millions of miles away, and to be in instant audio and visual communication with millions of our fellow men. We can perform calculations and solve problems that were impossible just a few years ago. But the new technologies also are jeopardizing privacy and substituting an information-processing capacity that has no ethical value system for a personalized capacity that does.

Taken altogether, these four advances bring within reach a human capacity to double, over a few decades, the *per capita* capability to transform the natural resources of the earth into the goods and services necessary to sustain life and to give meaning to sheer existence. Thus, the prospects for a better life for all in the near-term future are enormously brightened. On the other hand, together with the doubling of the necessities of life, there is likely to be a proportional increase in the world's population. There exists the possibility that the "carrying capacity" of the earth may be exceeded, with tragic consequences for the human race.

Whether society takes the road that will enable each individual to find self-fulfillment in a harmonious relationship with his fellow man and with nature or, alternatively, chooses the path that leads to a successively escalating series of catastrophes of famine, misery, terror, inequitable sharing of diminishing natural resources, and violent conflict between ideologies and between rich and poor will not be decided by some grand world plebiscite but will be decided in the manner indicated by John von Neumann 20 years ago. In answering his own question, "Can we survive technology?" he said, "yes, probably, provided that there is a long sequence of small but correct decisions . . . the intelligent exercise of day-to-day judgment."*

KEY SOCIETAL ISSUES

In the spirit of von Neumann, it may be helpful to express the question of the vitality of the human species in the form of a set of tractable societal issues, so that they can rationally be related to the atmospheric sciences. Without attempting to be exhaustive, it will be necessary to address effectively the following issues, some of which are related to the atmospheric sciences, if ultimate human disaster is to be avoided:

1. A population policy that will reconcile the imperative for a curtailment in the sheer numerical increase in the number of inhabitants on the earth with individual choices and the aspiration of those nations that can yet tolerate or receive more people.
2. A concerted effort to increase and stabilize food production so that the 60 percent of the present human population that is undernourished or inadequately fed will be brought up to a reasonable standard.
3. An overall increase in economic productivity, since this is the most formidable mechanism for increasing the *per capita* rate of converting natural resources into goods and services in the less developed countries. The more than tenfold gap between the *per capita* earnings of two thirds of the world's population in the developing countries and the remaining one third in the developed countries is widening.
4. Access to the natural resources in a manner that provides equity to each of the present and future inhabitants of earth. The current concern over energy is a case in point. One should take heart, however, in the Declaration of Principles adopted at the United Nations' Conference on Human Environment at which more than a hundred nations achieved unanimity on the right of each individual to share equitably in the world's natural resources. The fact that moral imperatives enunciated 2500 years ago have only now reached the stage of political agreement on principles—and remain yet to be implemented—may properly be the basis for a sense of urgency but should not be the source of dismay.

*J. von Neumann (1955). Can We Survive Technology? *Fortune 51*(6), 106–108, 151–152.

5. The livability of human settlements. The process of urbanization appears to be irreversible, but in some cases the disadvantages of aggregation of family units in an urban environment seem to outweigh the advantages.

6. An adequate *quality* in the physical environment. This involves the health implications, the economic impact, and the aesthetic considerations that enter into the management of air and water quality and land use. These matters involve decisions in the face of uncertainty and all the other elements of a multivariate problem with which the professional meteorologist is familiar.

7. Security of life and property in the face of natural hazards. As the world's population density increases, accompanied by an increase in the capital value of property, this issue becomes more and more important. A single two-day outbreak of nearly 150 tornadoes in the midwestern United States in April 1974 accounted for over 300 deaths, over 5000 persons injured, and more than 25,000 families suffering losses totaling more than a half billion dollars. The death toll from Hurricane Camille in 1969 in Mississippi and southeastern Louisiana was about 250, with over 5000 homes destroyed, some 5000 cattle drowned, and economic losses in excess of $1 billion—in spite of timely warnings that probably saved as many as 50,000 lives. In November 1970, a violent tropical cyclone moved over Bangladesh and took more than a quarter of a million lives.

8. The possibility of a massive use of nuclear weapons that, in a matter of minutes, could set back immeasurably the forward progress of man.

Given this list of societal issues, what can be said about the relevancy of the atmospheric sciences and about the knowledge base that should contribute to the "correctness" of that "long sequence of small decisions" that von Neumann held out as the probable manner by which we might survive.

NATURE AND MYSTERIES OF THE ATMOSPHERE

Consider for a few moments the earth's atmosphere—its nature, mysteries, and promises. We have come to appreciate that it is, physically, a seemingly thin and fragile—yet remarkably resilient—film of gases that encapsulate the planet and embraces within its lowest layers most of earth's biomass. The atmospheric envelope rotates with the globe but does not rest quietly upon it. The internal motions of the atmosphere—the winds—are in fact small departures from rather large absolute velocities measured with respect to outer space. Winds are induced by a nonuniform distribution of energy sources and sinks, which in many cases are formed and dissipated at various rates by the internal motions themselves. Air circulation patterns in the atmosphere range in scale from the currents circling the earth at high latitudes at altitudes from 20,000 to 40,000 feet (6 to 12 kilometers) down through the migratory cyclones and anticyclones, hurricanes, tornadoes, and thunderstorms; on into the small convective clouds; and finally into small-scale turbulence. One of the central problems of meteorology is to explain the existence of this wide range of atmospheric disturbances and the interactions among them.

In order to understand air motions, it is essential to know about the distribution of energy. The sun is the source of energy that drives atmospheric and oceanic currents as well. Incoming solar radiation is essentially balanced by outgoing infrared radiation from the land, sea, and air. Large quantities of heat are transported by the winds and by ocean currents. Theoretical and observational tools now exist to acquire a great deal of important information about radiative transfer of energy through the atmosphere. As a result of the internationally managed Global Atmospheric Research Program, scientists are learning about the energy exchange processes between the atmosphere and the ocean (which dominate the surface of the earth). As knowledge of the atmospheric energy budget is placed on a more secure foundation, we will make greater headway toward a real understanding of air motions or their consequences on the weather.

In addition to being thermally active, the atmosphere is the seat of many physical, chemical, and biological processes of importance. The chemical equilibrium of ozone in the upper atmosphere has a profound influence on biological processes on the ground as well as being a possible link between fluctuations in the emission of ultraviolet energy from the sun and atmospheric motion at lower levels. The distribution and decay of natural and artificial radioisotopes; the carbon, oxygen, phosphorus, sulfur, nitrogen, and other chemical cycles in the atmosphere; and the processes by which particulate materials are diffused and removed from the atmosphere are problems that have assumed ecological significance. The phase changes of water substances and the growth of cloud particles to raindrop size offer physical–chemical problems that are clearly of practical as well as scientific interest. The mechanisms of electric charge separation in a thunderstorm and the consequent lightning require further elucidation. The characteristics of

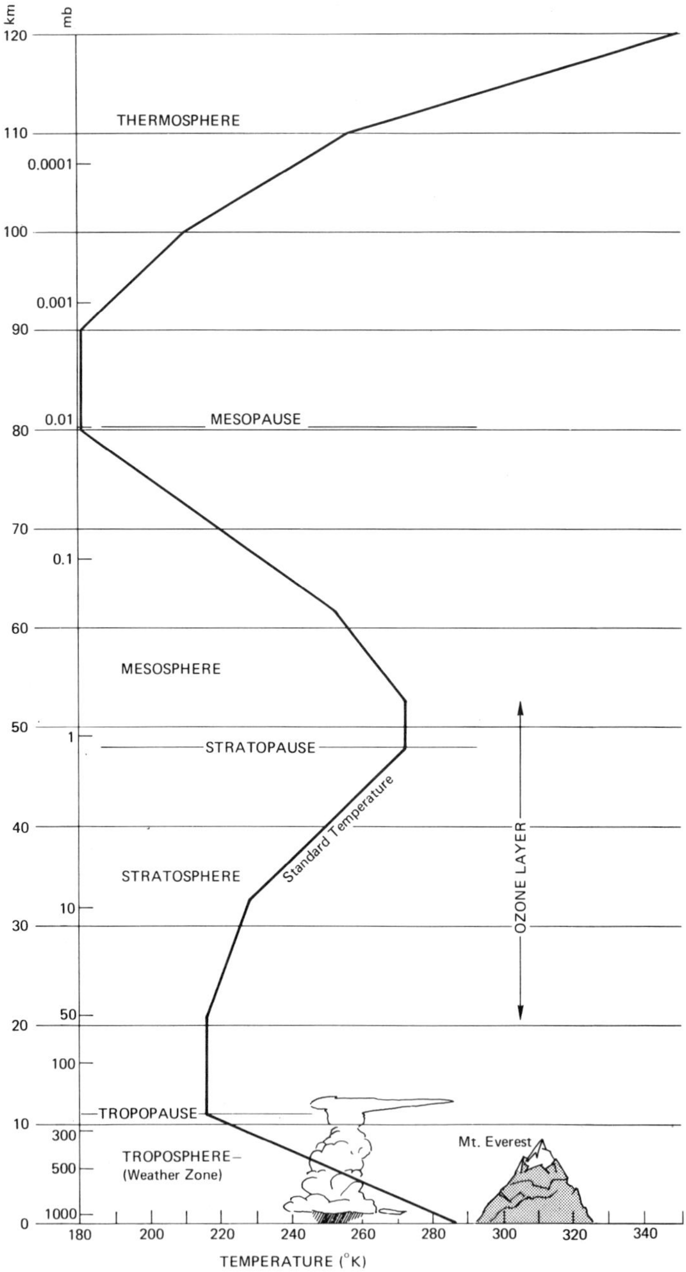

The structure and thermal characteristics of the standard U.S. atmosphere up to the thermosphere, beginning about 50 miles above the surface of the earth.

the atmosphere that influence the propagation of electromagnetic radiation over the entire spectrum of wavelengths are of obvious practical significance and of scientific interest in their own right.

Here in barest outline form is the nature of the atmospheric sciences: air motions; atmospheric energetics; and physical, chemical, and biological processes. These topics will be dealt with in greater detail in subsequent chapters of this report.

FUNDAMENTAL QUESTIONS

Much progress has been recorded since the end of World War II; much remains to be done. For present purposes, it is appropriate to identify certain questions of a fundamental nature:

- What are the limits of predictability in the atmosphere in terms of both the specific details of day-by-day weather and in the time—and space—aggregated atmospheric variable that we designate as the climate? Predictability may be viewed in either deterministic terms or in probabilistic terms.
- What are the factors governing seasonal variations in weather, fluctuations of global climate over periods of decades and centuries, and climate changes over periods of millenia and longer?
- What is the degree to which human activity can influence the behavior of the atmosphere either through conscious intervention in physical processes or as an inadvertent side effect of activities directed toward other objectives?

What are the prospects of obtaining answers to the large subset of scientific questions that will have to be answered to resolve these three basic issues? Very good! For several reasons:

- Enough is known about the physics and chemistry of atmospheric processes and related biological processes that many questions can be cast in classical quantitative, mathematical terms.
- The introduction of electronic computers into meteorological work over the past several decades has enormously multiplied the capability of handling the complex equations of atmospheric motion. The computing facilities available today provide a means for testing theory against observations.
- There has been an explosion in our capacity to measure important atmospheric quantities, from the global overview provided by polar-orbiting and geosynchronous satellites to the detailed three-

dimensional field of motions within a single convective cloud by means of a dual Doppler radar system.

• The magnitude of the scientific effort in the United States has increased severalfold since 1956 when Francis Reichelderfer, Chief of the U.S. Weather Bureau, asked the National Academy of Sciences to convene a committee on meteorology (now the Committee on Atmospheric Sciences) to propose measures that would accelerate progress in this field. The momentum generated by the activities of that committee is still increasing.

APPLICATIONS TO MAJOR NEEDS OF SOCIETY

It is necessary to examine the contribution that might be expected from the atmospheric sciences toward the resolution of the eight major issues confronting contemporary society that were listed earlier.

The matters of population and food will be considered together, since the most compelling argument for new population policies is the question of how 75 million additional people each year will be properly nourished in view of the fact that only 40 percent of the world's population is presently being fed adequately. The comment of A. H. Boerma, the Director General of the Food and Agricultural Organization in February 1973 is germane:

> One thing that has been harshly, even humiliatingly, made clear in the last two years is that, despite all our technological progress, despite all the buoyant hopes invested not long ago in the so-called Green Revolution, harvests are still far too often at the mercy of the weather. In this respect, at least, man has so far failed to master his natural environment.

He continued,

> . . . in the name of reason, can this world of the 1970's, with all its scientific prowess and its slowly growing sense of common purpose, go on enduring a situation in which the chances of enough decent food for millions of human beings may simply depend on the whims of one year's weather? Is this a tolerable human condition? Emphatically not!*

Clearly, it is essential to learn a great deal about the nature of seasonal weather and climate in order to develop adequate techniques for predicting year-to-year and longer trends in the statistical properties of the general atmospheric circulation. International agreement on this objective as one element in the Global Atmospheric

*J. E. Newman *et al.* (1974). World Climate and Food Supply, *Science 186*: 877–881.

Research Program (GARP) was reached at a meeting jointly sponsored by the World Meteorological Organization and the International Council of Scientific Unions in Stockholm in 1967. A second conference of 70 scientists from many parts of the world in 1974 led to the development of an action plan for an observational program and the development of improved mathematical climate models.

Since the international climate modeling program cannot be expected to produce results before a number of decades, a program to provide an "early warning" of reduced yields caused by climatic impacts was initiated by the World Meteorological Organization at its quadrennial Congress in 1975. This should permit more efficient management and planning of domestic international food supplies.

Available evidence indicates that, under appropriate meteorological circumstances, the weather may be changed in a favorable direction by cloud seeding or other weather-modification technology. More rain or snow, or less hail, could lead to substantial increases in agricultural production. The status and prospects of weather modification are discussed in a later chapter.

Substantial increases in global supplies of food and fiber could result from a more efficient use of available water and solar energy and by finding effective means of mitigating the stresses imposed by droughts, frosts, winds, and other weather elements. By improving photosynthetic efficiency in plants, greater yields would result. It has been proposed that increasing global carbon dioxide concentration may already be having such an effect. The combined efforts of agricultural biologists and meteorologists are needed to deal with these problems.

With respect to economic productivity, it is clear from various analyses that improvements in the precision and utilization of weather forecasts, as well as the extension of meaningful predictions up to, say, a week, would contribute substantially to the effectiveness of weather-sensitive commerce and industry. The development of more accurate one-week forecasts is one of the major goals of GARP, the most complex and highly integrated international scientific program ever conceived. Seven meteorological satellites, two in polar orbit and five geostationary, will interact with ocean buoys, constant-level balloons, and special observing stations to provide the initial-state data for the integration on very high-speed computers of the system of nonlinear partial differential equations that relate air motion to the thermodynamic conditions in the global atmosphere. Current analyses suggest that five- to ten-day predictions may constitute the upper limit in the deterministic predictions of daily weather.

It seems to be clear that the world is on the verge of adopting new procedures of natural-resource development and distribution—the fourth of the major societal issues that we listed earlier. For present purposes, it is convenient to divide national resources into three, somewhat arbitrary, major categories: (a) nonfuel minerals, (b) energy resources, and (c) renewable resources (principally food and fiber). The classes are not mutually independent. Production of aluminum, for example, is dependent on energy. For another example, one of the most copious supplies of energy in the United States is the low-sulfur coal in the Dakotas and Montana and the oil shale deposits in Colorado. Yet, unless special precautions are taken, extraction of these fuels could jeopardize the fragile land and water resources that are the sources of both food and fiber.

Given the fact that the natural resources of the world are finite, the characteristics of the era into which the world is moving with respect to the natural resources are twofold: (a) steadily increasing aggregate demand would, for at least many decades, lead to recourse to successively poorer stocks, greater expense, and "exhaustion" or substitution (any of which can be modulated by technology or conservation practices) and (b) geographical imbalances in supply and demand.

The relevance of the atmospheric sciences arises from the following considerations:

- More and more, man is turning to the marine environment, which is inextricably linked to the atmosphere. There has been a major expansion in the exploitation of oil, gas, sulfur, magnesium, bromine, and sand within the outer continental shelf. It is expected that by 1980 there will be more than a tenfold increase in the number of offshore oil wells. The technology has now been developed for the exploitation of the virtually unlimited deep-sea floor nodules containing manganese, copper, cobalt, and nickel. Most of these operations are highly weather-sensitive and place demanding requirements on weather forecasting and climatological analysis.
- The expanding world trade in food, fiber, minerals, and energy is going to multiply severalfold. The need for combined sea-state and weather predictions for the most effective routing of cargo vessels will increase accordingly.
- Promising results in experiments to enhance precipitation from winter storms in southern California and the mountainous far west provide the possibility for augmenting the hydroelectric power and other water needs connected with agriculture and mineral extraction in those areas. A troublesome aspect of this development was the reduction of federal government support in fiscal years 1974 and 1975 for

Large environmental (disk) buoy being prepared for anchoring in deep water off the coast of Alaska. (Courtesy, NOAA)

research and development in weather modification at precisely the moment when the prospects appeared to be most encouraging.

• The massive shift from the utilization of oil and gas for energy generation in the United States to the use of coal over the next few decades will exacerbate the potential health and economic risk from the sulfuric acid and suspended particulate sulfates derived from sulfur oxide emissions that combine with other pollutants, as well as that of the nitric acids and nitrates derived from nitrogen oxide emissions. Already, in some parts of the world downwind from large industrialized regions, rainfall acidity has increased. Tall smokestacks and various pollution-control procedures can and should be used to reduce toxicity and economic penalties. An important consideration, however, is the extent to which the atmosphere can be used as a common property of society to transport and dilute pollutants. To seek a return to the pristine conditions of a few centuries ago is unrealistic. To proceed on the assumption that there are no toxic, economic, or aesthetic penalties is equally unrealistic. It is necessary to use the dispersive powers of the

Oil drilling platform in North Sea buffeted by heavy seas. (Courtesy, Asiatic Petroleum Corporation)

atmosphere without increasing morbidity and mortality. Much progress has been made in recent years in both the modeling of the dispersive power of the atmosphere and in epidemiological studies to believe that rational solutions to complex problems can be achieved over the next decade.

The recognition that human activity can significantly influence weather and climate is of relatively recent origin. It is well known, for example, that within cities certain climatic elements differ significantly from those in the open countryside. It is possible that man's influence on the atmosphere may become more profound during the next century

Oil transport tanker in stormy seas. (Courtesy, Exxon Corporation)

or two, if population and *per capita* consumption of energy continue to grow at anything like the current rates. If the energy dissipated into the atmosphere over areas of 10^6 km^2 (1 million square kilometers) exceeds a few percent of the net solar radiation, it is not inconceivable that meaningful climatic changes could occur.

Notable progress has been made during the past five years in quantitative assessment of the local and regional impact of urbanization on the atmosphere through the intensive activities of a multi-institutional research program in the St. Louis area. Supported by several governmental agencies, the goals of the program are fourfold: (a) to identify the effects of urbanization on the frequency intensity and duration of various aspects of atmospheric behavior, among them precipitation production and related severe weather; (b) to identify the physical processes of the atmosphere through which the observed effects are produced; (c) to identify the factors of urbanization that are causative agents of the observed effects, with a view to the possibilities for purposeful manipulation of these factors in the future; and (d) to begin assessment of the present-day nature and magnitude of the impact of these modifications of atmospheric behavior on the wider environment-related issues of society. Results to date suggest a 10 to 15 percent increase in summer precipitation. Hail and thunderstorm intensity increased sharply downwind. There is a marked heat island effect over the urban area itself, and the low-level airflow is markedly perturbed. The increased rainfall adds roughly 15 percent to the local streamflow.

Local crop yields are estimated to be augmented by 400,000 bushels (about 10,850 metric tons) of corn and 130,000 bushels (about 3500 metric tons) of soybeans—more than enough to offset the increase in hail damage. Conceptual models are now being developed relating the local and regional climate to urban-industrial activities, and the groundwork is being laid for an assessment of the overall environmental impact and exploration of the possibilities of purposeful intervention in the urban activities to optimize the impact. Such information will be useful in planning and building new urban areas and redeveloping old ones.

Enough has been said already about the importance of the atmospheric sciences to the quality of the local physical environment, but two more matters of worldwide concern deserve mention.

- The first is the possible impact of a severalfold increase in the carbon dioxide (CO_2) accumulated in the atmosphere as the result of the combustion of fossil fuels. The increase during this century is in the 10 to 15 percent range, and this figure would probably be twice as high were it not for the buffering by the oceans. Very preliminary calculations of the consequences of a doubling in the atmospheric CO_2 suggest that the global average surface temperature might rise up to 2°C, a climatically important amount. The global circulation model from which this calculation was derived is oversimplified in that it does not take into account interaction of clouds and their effects on the heat balance and the heat storage capacity in the continents and the ocean, nor the transport by the oceans. More realistic models are being developed to investigate this important matter.

- A second human-cause perturbation having potential global implications that is now under serious study is the possible consequences on the thin protective layer of ozone by injecting into the stratosphere the oxides of nitrogen from supersonic aircraft or the diffusion of inert chlorofluoromethanes into the stratosphere from ground levels, where they are used in refrigeration systems and as a propellant in household aerosols. Chlorine atoms are produced through photodissociation by ultraviolet radiation and attack ozone at altitudes above 25 km. Oxides of nitrogen and chlorine in the stratosphere result in chains of chemical reactions that may have the net effects of reducing ozone concentra-

Industrial pollution in the eastern United States. (Courtesy, NOAA)

tion, thereby permitting more ultraviolet light to penetrate to the biosphere. Increases in ultraviolet radiation reaching the ground can be expected to increase the incidence of skin cancer in humans and affect plant and animal life deleteriously. While uncertainties surround both the atmospheric and biological processes as well as the degree of hazard involved, the matter deserves the serious attention it is now receiving.

Hurricanes, lightning, and tornadoes caused annual property damage that averaged $500 million, $200 million, and $150 million, respectively, during the period from 1955 to 1970, and exacted an annual toll of lives that averaged 75, 150, and 100, respectively, over the same span of years. A major program, known as Project Stormfury, has been under way for a number of years, with the objectives of developing an understanding of the genesis, structure, and behavior of these hurricanes and of testing if cloud-seeding techniques can reduce the peak wind speeds in hurricanes. During a "stand-down" phase since 1972, while aircraft instrumentation was being modernized, it was hoped the program could be reactivated in the Pacific, where the frequency of hurricanes is three times greater than in the Atlantic Ocean. Unfortunately, objections to hurricane seeding by Japan and the Peoples' Republic of China have aborted those hopes. In 1976, plans were being made to resume tests in the Atlantic area. The preliminary results in the Atlantic, and the value of being able to ameliorate the ravages of hurricanes and typhoons, represent strong arguments in favor of an early start for this program.

Planned for future implementation is the Severe Environmental Storm and Mesoscale Experiment (SESAME), with the objective of clarifying the genesis of severe storms and the identification and quantitative evaluation of the mechanisms that locate, trigger, and modulate severe storms. Closely allied with this effort are plans to model in three dimensions a single deep convective cell in the atmosphere.

In the foregoing paragraphs, we have touched on the contributions that the atmospheric sciences can make to several of the crucial issues confronting society today. Taken together, they would seem to provide a reasonable justification for the dollar or so per person that annually goes to support research in the atmospheric sciences in the United States, as well as the two to three dollars per person per year that underwrite the total meteorological services.

In considering the eighth, and final, issue confronting contemporary society—obliterative warfare—the reasoning changes subtly, yet profoundly. The interdependence of "national" atmospheres has been recognized for over a century. There is probably no specialized agency of the United Nations that is more tightly interwoven—and it might be added, more effective—than the intergovernmental World Meteorological Organization (WMO). Nor is there any nongovernmental organization more "one-world" oriented than the International Association of Meteorology and Atmospheric Physics, a subsidiary body of the International Council of Scientific Unions (ICSU). Even the intergovernmental organizations and the nongovernmental organizations have found a common cause in the support of the Joint Organizing Committee, established by WMO and ICSU to direct and oversee the activities of the Global Atmospheric Research Program (GARP). The reason is clear. The atmosphere is no respecter of national boundaries. It is, in the fullest sense of the term, an "international common property" and resource. It is not feasible to attempt predictions over one nation without a sensitive awareness of the initial conditions over all nations. Nor can alterations, either conscious or inadvertent, be made over one country without affecting the atmosphere over other countries. The fabric of international cooperation in the collection and exchange of data and in basic and applied research is a model that other fields might well emulate.

No more demonstrable or vivid example of international interpreta-

Artist's view of the synchronous meteorological satellite (SMS) positioned over the equator at about 22,000 miles altitude. (Courtesy, NASA)

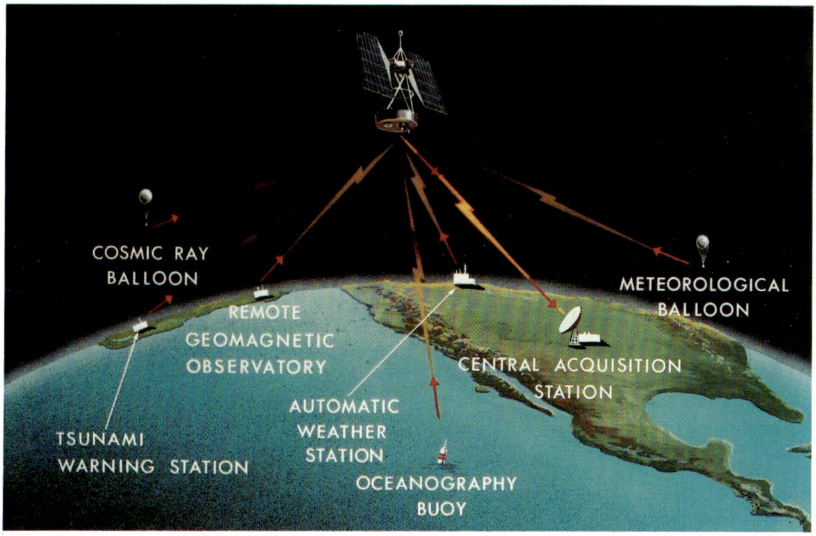

A satellite data collection system showing the research and developmental NIMBUS satellite of NASA. (Courtesy, NASA)

Center-Pivot irrigation systems altering the landscape and hydrologic relations in north central Nebraska. (Courtesy, University of Nebraska)

A small windbreak of corn plants sheltering sugar beets. (Courtesy, University of Nebraska)

The modern research/survey vessel, *Oceanographer,* which participated in the Atlantic Tropical Experiment out of Dakar during 1974. (Courtesy, NOAA)

The recently acquired research aircraft of the National Oceanic and Atmospheric Administration's Research Facilities Center, located at Miami, Florida. (Courtesy, NOAA)

tion in science can be cited than the GARP Atlantic Tropical Experiment (GATE), which brought together during a 100-day period, in the summer of 1974, 4000 scientists and technicians; involved 70 nations; and employed 40 ships, 13 aircraft, numerous ocean buoys, other special observing equipment, two polar-orbiting satellites and one geosynchronous satellite, and a vast array of conventional surface observations in an intensive and highly synchronized field program over an area of 500,000 square kilometers in the eastern tropical Atlantic. This program is leading to a better understanding of the interaction among four different scales of motion in the tropics, ranging in size from 1–10 km up to 200–10,000 km. Such information is essential for the development of the mathematical models of the global atmosphere.

Good scientific research is needed to solve critical societal problems, and the urgency of the problems can provide the stimulus for good research.

2

Impacts of Weather and Climate

CLIMATIC IMPACTS ON WATER, FOOD, AND ENERGY

When the world's population was small, it appeared that the earth had an inexhaustible supply of water, food, and energy. People sometimes had to labor hard to obtain the necessary ingredients for life, but few gave much thought to notions of limited supplies. In the meantime, the population continued to increase at faster and faster rates. In recent years, the number of humans on this planet pushed past the four billion mark and promises to be seven billion in three decades. In the subsequent few decades the number might double, assuming there is no major catastrophe involving water, food, or energy.

We have already seen widespread starvation and malnutrition in the developing countries. Even if the weather were beneficient in the years to come, would there be enough food and water for these additional billions? There are reasons for believing that the advanced countries will keep their populations under control and produce adequate food to sustain good health, but there are grave doubts about the welfare of a large fraction of the world's population.

Agricultural meteorologists note that general climatic conditions during this century have been unusually good in comparison with conditions over the preceding centuries. The gradual cooling of the northern hemisphere since about 1940 has been documented and is seen by some climatologists as the start of a long cooling period. Its continuation during the past decade and the southern hemisphere trend even since 1940 are questionable. Nevertheless, it is not necessary to resort to predictions of ice ages to cause grave concern. Cooling of only a degree or two on a worldwide basis could have profound effects on the general characteristics of the storm patterns. It could lead to reduction of the length of the growing season in many parts of the world. It could conceivably lead to the spreading of deserts and the change of monsoon circulations.

Even if the long-term climate of the globe were to undergo only minor changes over the next few decades, we can be certain that there will be pronounced short-term fluctuations in the climate. As in the past, droughts will occur. The Soviet Union saw some major ones in 1972 and 1975, and the social consequences were serious. Fortunately, over the central part of the United States and Canada, truly the breadbasket of the world, the weather has been reasonably favorable. Notwithstanding the serious Great Plains droughts from 1954 to 1957, in 1974 and 1976, and the 1963 drought over the eastern states, North American farmlands have not seen widespread, persistent drought such as those of the 1930's. Anyone familiar with climatic records is convinced that droughts will recur, but no one can say for sure whether it will be this year, next year, or some year farther in the future. Some scientists have related weather conditions with sunspot frequency or other solar events and are prone to make predictions of seasonal weather. Past experience shows that they have not been particularly skillful.

Climate and Water*

Water is critical to every facet of life. The water resources of a region or a nation control its ultimate activities, whether it be the urban dweller who is affected by his water supply and sanitary disposal system or the farmer and his crop production.

Climate has often been the determining factor in setting the stage for the water resources found in the region. Consider the Great Lakes region of the United States. Before the Pleistocene Epoch, the area where the Great Lakes exists was a broad, flat plain. The cooling of the climate of the northern hemisphere during the Pleistocene, which eventually involved four massive glacial stages that partially covered

*The major portion of this section was contributed by Stanley A. Changnon, Jr., Illinois State Water Survey.

North America during different periods, produced the Great Lakes and all other sources of water in that area of North America. The oscillations of the glaciers gouged out the lake basins, and the meltwater filled them. The oscillations of the glaciers over the land portion of the Great Lakes basins also deposited vast layers of sands and gravels, which now contain much of the groundwater of the area. The glaciers also formed hundreds of small lakes throughout the Great Lakes region, adding to the surface water supplies. The warming trend of the earth's climate in the last part of the Pleistocene and subsequent wind and water erosion have been responsible for the present shape of the Great Lakes.

An understanding of the water resources of an area, whether it be those of a county, a state, or a nation, comes easiest by thinking of the hydrologic cycle and its four key phases. First, there is the basic input—the precipitation. In the second phase, there is the infiltration of the water into the ground, some of which becomes stored groundwater, often moving laterally in different layers below the surface of the earth. A third component of the hydrologic cycle is the runoff, that is, the movement of water on the surface. It may be as overland flow or as streamflow that runoff eventually returns to the oceans. Runoff in the form of streamflow varies over a particular area in direct relationship to the amount of precipitation. Finally, we complete the cycle by the return of moisture from the land and open water surfaces to the atmosphere. This involves evaporation from land and water surfaces and transpiration of moisture from plants to the atmosphere.

Every phase of the hydrologic cycle is strongly controlled by the weather and climate. The degree of moisture in the atmosphere at any time, coupled with the amount of instability, often produced by contrasting air masses, determines the occurrence, place, and amount of precipitation. Large shallow storms of the cold season in North America produce rains or snows over broad areas. Less widespread but more unstable fast-moving fronts of the warm season induce showers and storms that produce rainfall over limited areas. Clearly, the input phase of the hydrologic cycle is controlled by the atmospheric processes. The infiltration of groundwater and surface runoff, the second and third phases, also are controlled largely by climatic factors. Cold temperatures freeze the ground and restrict the movement of water into the soil layers. This also increases the runoff, and the amount of runoff can be drastically affected by thawing temperatures, which produce rapid snowmelt in the northern latitudes. Clearly, temperature is a major factor in controlling the motion and distribution of water at and near the surface of the earth.

Finally, the fourth phase of the hydrologic cycle, evaporation and transpiration, is also greatly affected by climate factors. Solar radiation, air temperature, the amount of water vapor in the air, and winds all interact to determine the rate and hence the amount of moisture extracted from the earth.

Climate is also important in another key aspect of water resources—water quality. Pollutants emitted into the atmosphere are transported varying distances, depending on the vertical mixing and the wind velocities. Some pollutants are converted through the action of sunlight, water vapor, and cloud processes into substances harmful to man. A substantial fraction of the air pollutants is washed out of the atmosphere by rainfall. Large quantities of the contaminating substances found in the bottom sediments of the Great Lakes came from airborne material deposited into the lakes, rather than from water pollutants carried into the lakes by streams and rivers.

A good image of the relative impact of weather and climatic factors on water resources in an area can be illustrated by considering the quantities of water available for Illinois, a typical midwestern state. The great inflow of atmospheric moisture, or water vapor, in the air averages 6.2 million acre-feet (7.6 billion cubic meters) per day. From this enormous source of moisture passing over the Midwest, the rather inefficient processes of the atmosphere cause only about 5 percent of this total to fall as rain or snow. However, this 5 percent is an average of 305,000 acre-feet (1 acre-foot is approximately equal to 327,000 gallons) per day for Illinois. Evaporation from land and water surfaces and transpiration from growing plants have first call on this water. Together these processes consume and return to the atmosphere about 230,000 acre-feet per day, or 75 percent of the amount that falls as precipitation. The remaining 75,000 acre-feet per day left in Illinois runs off as streamflow or infiltrates as groundwater. This amount, when added to the minimum flows of the Mississippi and Ohio Rivers, as well as the pumpage of water from Lake Michigan, brings about 132,000 acre-feet per day, the grand total water supply available for use or storage in Illinois. This is an immense amount of water—five times the present state usage and one sixth of the water usage for all purposes in the entire United States. This explains why Illinois and most of the states east of the one hundredth meridian are "water surplus states."

Although at least half the area of the United States, on the average, has water surpluses, climatic fluctuations that bring severe droughts can lead to deficient water supplies. It also must be realized that water is not uniformly available in place, time, or quality. High-demand regions, such as major cities, can overtax available supplies, particularly in times of drought.

Weather fluctuations also affect the usage of water resources. Periods of high temperatures in urban–industrial complexes lead to increased water usage for cooling purposes. Exceptionally low temperatures can lead to the freezing of major transportation arteries, such as the upper Mississippi River and the Great Lakes. Abnormally high temperatures greatly affect the rate of water usage for irrigation.

In summary, the water resources are one of the major controlling and limiting factors for the flora and fauna of a region. Water sets limits on crop production, on urban and industrial development, and on the ultimate size of the population that can be supported. In turn, climatic fluctuations of the past have often played crucial roles in developing the basic water resources now available in a region. Furthermore, the fluctuations in weather of any given day, month, or year determine the flow of water through the hydrologic cycle. A lack of knowledge about the major climatic conditions—precipitation, temperature, atmospheric moisture, and winds—greatly limits the understanding of the water resources of a region. The wise management of these water resources, including the development of reservoirs, the rate of groundwater pumpage, the growth and placement of urban and industrial activities, the diversion of water for irrigation, and the density of crop plantings should be rooted in climatic data and knowledge of the limits set by the climatic variations.

Climate and Food Production*

Man's ability to produce food for more than just his own family has improved steadily since the beginning of the industrial revolution. Agricultural productivity has increased at a more or less constant rate in developed countries during the twentieth century. This increasing productivity in much of the industrialized world gave rise to assumptions that science and technology had the capacity to provide sustained, and ever-increasing, food supplies. We have tended to ignore or forget that, in virtually every year of recorded history, food shortages have been serious enough, somewhere in the world, to lead to loss of human life.

Many less-developed nations have made valiant efforts to achieve self-sufficiency in food production. Because population growth has proceeded at a greater rate than the increase in food production, most developing countries are less well off today than they were 20 and 30 years ago. Nevertheless, only two or three western nations can be considered reliable net exporters of food today. All others have be-

*The major portion of this section was contributed by Norman J. Rosenberg, University of Nebraska.

come dependent, in some measure, on the stable production of these few nations. And this has occurred notwithstanding the fact that weather has been generally good in important food-producing areas of the world such as the prairies and Great Plains of North America since the 1940's. Despite this recent history of good weather, world granaries now hold enough food to feed the world's people for no more than about 60 days.

National weather services have been developed, primarily, to provide support for aviation, transportation, and other industries and to alert the public where and when its safety is threatened. The needs of agriculture have received relatively little attention, and this fact is reflected in the budgets of most national weather services. This neglect may stem, in part, from the fact that agriculture is a complicated business. While the needs of aviation are relatively easy to define, the impacts of weather and climate on agriculture are difficult to quantify, except where they are overpoweringly evident, i.e., frost, hail, drought, and windstorm.

Climate Changes and Fluctuations Agricultural changes by adaptation occur through a long process of experience in which a particular species or variety proves itself to be the most reliable producer.

As discussed in a later chapter, there is considerable speculation on the prospect of future global climatic changes. Since they are a part of earth's climatic history, it is surely reasonable that we consider and review the possible impacts of continued change and fluctuation.

A cooling of the northern hemisphere climate, for example, might shorten the growing season for many temperate-zone crops. The growing region for our major grain crops might be shifted southward and might also be reduced in size. Given enough warning, however, the agricultural research establishment throughout the United States and much of the developed world is capable of reacting to predictable long-term shifts in climate through the introduction of new, adapted varieties. Unfortunately, such shifts cannot be predicted, and the prospects for developing such predictive capability over the next few decades are not encouraging.

The prospect of a slowly changing climate may be less ominous, however, than the prospect of increased interseasonal and intraseasonal variability. It is unlikely that any plant variety can be bred to produce best at both hot and cold temperatures or under both wet and dry conditions or to recover quickly from every type of stress imposed by extremes in a fluctuating climatic situation. Indeed, the varieties that produce best in the long term are those that are generally adapted to median "normal" climatic conditions. A major long-range goal for the meteorological profession in support of world agriculture is to provide reliable predictions of future climatic change and of the nature and degree of climatic variability. These tasks pose great difficulties. They will require major investments of scientific talent and funds over many decades.

Fluctuations in the availability of fertilizer, fuel, and seed and of disease- and insect-resistant varieties have significant impact on the stability of agricultural production. But weather is the dominant factor in determining global and regional food production from year to year.

Water Supplies and Agriculture The stability of agriculture and the potential to improve crop and animal production worldwide depend, in a crucial way, on our ability to stabilize and improve the supply of water and the efficiency with which it is used. Micrometeorologists, plant physiologists, and soil physicists have already learned much about the water requirements of major world crops. Accurate measurements of evapotranspiration over natural surfaces have been made with instrumental methods based on energy balance and aerodynamic theory. Empirical methods have been developed for predicting the potential evapotranspiration or water demand—that which is determined by climatic conditions when water supply to the growing plant is unlimited. The actual evapotranspiration—that which occurs when soil and plant factors exert a resistance to the transfer of vapor from soil into the air—is less well predicted by available methods. Despite great efforts, acceptable empirical methods for predicting the actual evapotranspiration component of the hydrologic balance still elude us.

Recent research has shown that the surface albedo, or reflectivity, of soils changes systematically with the water content of the soil. Soil-surface temperature and soil thermal inertia indicate whether a soil surface is wet or dry, and crop canopy temperature indicates whether the crop is transpiring water freely or is in a wilting state. Thus, improved methods of remotely sensing albedo and surface temperature of large units of land may, in the future, provide better methods for estimating actual evapotranspiration.

Water storage and distribution systems depend on reasonable watershed-scale estimations of precipitation and runoff as supply factors and of evaporation and evapotranspiration as demand factors. There is need for improved climatological techniques to estimate precipitation probability, especially in the arid and semiarid zones where rainfall is most irregular and in the regions where growing season rainfall occurs as widely scattered thunderstorms.

The photosynthetic conversion of carbon into sugars and other more complex compounds requires energy. Solar radiation is the source of the energy stored in the photosynthate. Even under the best modern agricultural systems in use today, not more than about 2 percent of the solar energy that impinges on a field during a growing season is "harvested" in plant products. The known quantum efficiency of the photosynthetic reaction indicates that as much as 8 to 10 percent of the solar radiation should be harvestable. Thus, there is great potential for increasing global food production through an increase of photosynthetic capture of solar radiation.

In order for the plant to function well, its tissues must be hydrated and its temperature must remain within an optimum or, at least, a nonlethal range. A good water supply to the plant is essential if these conditions are to be met. Well-developed, healthy, and vigorous crops do not necessarily consume more water than do less well-developed crops. Thus, yield increases due to improvements in the photosynthetic process need not involve increases in the use of water. In fact, some production increases may be achieved with reduced water consumption, especially if irrigations are timely and rainfall in the non-growing season is effectively stored in the soil for later use. Water saved in one field can be diverted to introduce additional acres into cultivation. A major objective of agricultural meteorology is to increase the water use efficiency, which we define as dry matter produced by photosynthesis per unit of water consumed in evapotranspiration.

Frost and Freeze Damage Some of the world's most productive land is found in temperate regions, where cold weather limits the length of the crop-growing season.

Length of the growing season is determined, essentially, by the onset of relatively freeze-free weather in spring and the first occurrence of killing frost in fall. Climatological descriptions of growing season length have been based, largely, on statistics such as first 0°C temperature in fall. Since, conveniently, frost date series are randomly and normally distributed, probability statements on "mean frost date" are easily computed and widely used.

It is generally true that the final frosts in spring and the first frosts in fall are radiative types (involving ground-based temperature inversions) and that, in such frosts, protection is possible. The cost of protection must be weighed against the advantages in yield or market to be gained from an extended growing season.

Even in subtropical regions, where the statistical "growing season" is normally 365 days, an occasional frost may cause severe economic loss of high-value vegetable and fruit crops. The production of an entire season may be lost or valuable perennial orchards may be taken out of production for a few years or even be totally destroyed.

Accurate predictions of the degree and duration of frost conditions are essential, but such forecasts are of little value unless protective measures are possible. The measures that are most commonly used in high-value citrus and other fruit orchards involve the disruption of thermal inversions by means of powered wind machines or with orchard heating methods. Heating is accomplished by the combustion of various fuels to maintain the air within the orchard at a temperature above freezing.

Alternative means of frost protection that consume less fuel are needed. For example, it has been shown in a Texas experiment that heat loss from a grapefruit grove can be diminished by lowering the thermal emissivity of the leaves with a coating of an aluminum-containing material. A plastic screen has been used to surround and cover a fruit orchard in Idaho. During a frost, the screen is wetted by sprinklers and the resultant ice shield protects the enclosed vegetation from the cold external environment. Within the screen, sprinklers coat the plants with water, maintaining the plants and air at temperatures near 0°C. Foams have been used successfully to cover young plants during a frost night in a vegetable production area in Canada. Further development of such nonheating techniques will involve effective micrometeorological research. The adoption of such methods will rest on climatological studies of the frequency with which they can be successfully used.

Wind Damage Mechanical destruction of crops in severe weather is another constraint to stable agricultural production. Hurricanes and cyclones do considerable damage to crops grown in certain coastal regions. This aspect of their destructiveness is rarely commented on, in view of the greater drama attached to loss of life and loss of property.

Strong and damaging winds often reduce agricultural productivity inland as well. Cold winds in spring and fall may cause mechanical damage to the whole plant as well as freezing damage to certain tissues. Winds blowing from arid into semiarid and subhumid regions also cause mechanical damage. These winds, because of their high temperature and low humidity, impose severe moisture stress on the growing crops, as well, and cause wilting, desiccation, and the loss of potential productivity. In regions where the land is not well protected by vegetation, wind erosion may occur and initiate a decline in productivity. Young, tender vegetation may be damaged or destroyed by "sandblasting" when soil is eroded by wind.

Properly designed windbreaks can aid greatly in stabilizing agriculture in regions where strong winds are common. The windbreak aids in uniformly distributing snow over fields, thereby increasing the supply of soil moisture in spring. Windbreaks have a considerable impact on the crop they shelter during the growing season as well.

Considerable experimentation with tree windbreaks and with windbreaks constructed of such materials as snow fencing, plastic screens, and reed mats have shown that the climate that prevails in the sheltered area is more moderate than that in adjacent unsheltered fields. The air is slightly warmer by day and slightly cooler by night, but absolute humidity is greater by day and by night. The overall effect on the microclimate is such as to moderate evaporative demand and moisture stress on the sheltered plant. Since moisture stress leads to wilting, closure of the plant stomates, and cessation of photosynthetic activity, the windbreak should permit improved crop yields. Evidence from around the world shows this to be true.

Despite the proven beneficial effects of windbreaks planted in the Great Plains during the drought years of the 1930's, many of them are

Severe soil erosion by wind during the major drought of the 1930's. (Courtesy, U.S. Department of Agriculture)

now being removed. Changes in agricultural land use that involve larger fields and expensive irrigation systems have increased the value of the land to the point where farmers begrudge its occupation by tree windbreaks. Windbreaks may interfere with the mechanical operation of the large center-pivot sprinkling systems that are revolutionizing irrigation in the Great Plains region.

There is urgent need for windbreak designs that are compatible with current and forseeable agricultural systems in windswept regions.

Diseases and Insects Three conditions must be met before disease or insects can infest a crop: the disease organism or insect must be present, the host plant must be susceptible to attack by the parasite, and the environment must be conducive to activity and reproduction by the attacking organism. Literally thousands of diseases and insects that attack plants have been identified. There is a vast literature categorizing plant species and varietal susceptibility to diseases and insects. Unfortunately, the specific environmental conditions conducive to the proliferation and attack by a parasite are known in only a few cases.

Today the control of disease and insects is accomplished in agriculture largely by the application of chemical sprays and dusts. Increasingly sensitive to the introduction of chemical agents into the food we eat and the air we breathe, our society is becoming more critical of this approach. Methods other than chemical control or that involve limited use of chemicals for control of disease organisms and insects are now being developed.

An appropriate strategy for improved chemical control of pests may involve the following steps:

1. Define the environmental conditions conducive to disease or insect outbreaks, e.g., temperature, humidity, leaf wetness, windiness, illumination.
2. On the basis of weather records, determine the probabilities that these environmental conditions will prevail at any specific time and for any specified duration.
3. On the basis of up-to-date information, prepare weather forecasts indicating the probability of occurrence of conditions conducive to the disease. Only then would actual spray operations be considered.

Such a combination of biology, climatology, and meteorological forecasting has led to development of effective systems for control of the late-blight disease in potatoes in Ireland and is used in other potato-growing nations, as well.

Agrometeorologists may help in new ways to develop strategies other than chemical control of plant pests. The white-mold disease of beans provides an example. This disease flourishes in humid regions when the weather is cool and damp for protracted periods. The disease has also become important in semiarid regions where irrigation of bushy varieties creates a conducive microclimate. New designs of the crop-canopy architecture, which can be introduced by plant breeding, in conjunction with altered irrigation schedules are now under study. These adaptations may permit the maintenance of a microclimate less hospitable to the white-mold disease.

There has been other evidence of progress in applying meteorology to the control of plant disease. The spread of southern leaf blight of corn, for example, which threatened much of the 1970 crop in the corn belt, has been well modeled. An alert system was devised using a computer model in which observations on duration of leaf wetness, air temperature, and humidity are obtained from zones in which the crop is susceptible to the disease. Additional models of this kind can be developed, given adequate knowledge of the biological relationships and some understanding of the mechanisms by which spores or other vectors of disease are transported from zone to zone.

Animal Agriculture There are a number of opportunities for increased service by the atmospheric sciences in support of animal agriculture and grazing. The most immediate opportunities lie, probably, in the realm of forecasting. The animal has an advantage over the plant in that it is able, to some extent, to escape stressful conditions. A sheep standing in bright sunlight is able to seek out shade, whereas the plant by which it stands, lacking water, can only wilt. On the other hand, in the case of severe weather, e.g., floods in lowland pasture areas, blizzards on the open range, hailstorms, or extended drought over large regions, the domesticated animal often has no escape. Rapid and accurate forecasting and communication of advisories to the rancher or herder may enable the most timely action to protect against catastrophic losses of animals.

Meteorological information can also be helpful in regions of short forage supply by pinpointing the location and amounts of rainfall. In some of the least-developed countries, as in the Sahelian Zone of Africa, one of the few applications of meteorology now practical would be to guide herders to areas where rainfall makes pastorage available.

Concluding Remarks A better understanding of how weather and climate affect agriculture allows the development of procedures to optimize production. Accurate and timely weather forecasts for as long a period as possible have great value for most agricultural activities. Prediction of climatic conditions, either in probabilistic or deterministic terms, can be used for planning future food production and distribution strategies.

Interactions of Energy and Climate*

A number of the most significant interactions between man and his environment relate to the interaction between energy use and climate. Energy, of course, is fundamental to the technological societies of both today and tomorrow. Plausible projections of future world population and energy use suggest a twenty-first century world population of about 10 billion, with total energy requirements more than five times that of the present.

Such a crowded and energy-hungry world will clearly stress the finite resources of our planet and will have to develop a stable and secure *modus vivendi* with the natural environment. The uncontrollable and most changeable element of that environment is the climate, which we now realize varies significantly on all time scales.

Fluctuations in climate, whether a season or a decade in length, clearly influence man's need for energy in many ways. As noted earlier, climate is crucially important in agriculture, which is becoming increasingly dependent on energy for cultivation, transportation, and fertilizer production, for example. Year-to-year variations in seasonal weather clearly will influence the timing and magnitude of energy needs. Longer-term climatic variations might completely change the nature of agricultural production in large regions, thereby inducing massive shifts in the use of energy resources. Similarly, requirements for heating fuel, optimal transportation routing, and many other aspects of human society are influenced by climate. Rigorous optimization of all these subsystems of our complex world society will be essential if the needs of future world society are to be met. One can, therefore, expect that this optimized structure will be increasingly sensitive to variations in critical elements of its environment, such as climate. Since energy is fundamental to all of man's activities, one can foresee ever closer relationships between total energy needs and climate variations.

Another class of problems, with disquieting possibilities of poten-

*The major portion of this section was contributed by John S. Perry, National Research Council.

tially dangerous feedbacks, is posed by the threat of climate alterations induced by human actions, most of which are related to the consumption of energy. Environmental pollution of some sort is an inescapable consequence of energy use. At the very least, heat is inevitably dissipated in proportion to the energy consumed. But, perhaps surprisingly, a plausible scenario of future human development postulates that the vast preponderance of world energy requirements through the twenty-first century will continue to be met by combustion of fossil fuels, principally coal.

Burning of fuels for energy produces additional by-products besides heat. One may project that in the next century annual releases of heat and carbon dioxide will be eight or nine times current levels, while emissions of particles could be as much as twenty times current values. Such massive pollution is clearly cause for grave concern. All of these factors have potential impacts on climate.

Heat release is already significant on a local scale. Manhattan Island now releases about as much man-made heat as it receives from the sun, and all major cities are "heat islands" in which average temperatures are notably different than their surrounding countrysides. Through a combination of heat, moisture, and particle injections, cities modify climate for a moderate distance downwind. In the future, we can anticipate continued growth of urban centers and perhaps the consolidation of energy production into "power parks." These concentrated sources of heat may well have significant influences on downwind climate, and these effects may prove to be important factors in siting decisions. The climatic effects of regional energy concentrations are difficult to assess with present-day techniques, but one should expect climatic changes roughly proportional to the area and intensity of the heat release.

The role of particles in climate is complex indeed. On a global scale, man-made particles are now and will remain a relatively small contribution to the great mass of natural material suspended in the air. These particles are crucially important in the processes of cloud and precipitation formation. Moreover, particles influence the transfer of solar and heat radiation through the atmosphere. The net climatic effects of anthropogenic particles, however, depend on their nature and distribution within the atmosphere, factors difficult to forecast into the distant future.

Gases produced by combustion similarly affect the transfer of solar radiation. Of these, the most ubiquitous and important is carbon dioxide. It strongly absorbs and emits heat radiation in regions of the infrared spectrum through which the earth radiates its heat to space.

Thus, increases in carbon dioxide tend to insulate the earth from space and make it warmer. At present, about half of each year's release of carbon dioxide remains in the atmosphere; the remainder is taken up by plants or dissolved in the ocean. Since the Industrial Revolution, the concentration of carbon dioxide has increased by about 10 percent because of fossil-fuel combustion.

As we look to the future, all signs point to a continuing massive increase in carbon dioxide concentrations as population and living standards increase. As noted earlier, economic considerations suggest that the bulk of man's energy needs will continue to be met by burning of fossil fuels. Furthermore, careful analysis and modeling of the ocean's chemistry and circulation predict that the ocean's ability to absorb carbon dioxide will markedly decrease as more of the gas is dissolved. It has been projected that by the years A.D. 2150 to A.D. 2175, the peak atmospheric concentration of CO_2 might be five to seven times the preindustrial level, with concentrations well above present values *persisting for hundreds of years.**

The magnitude of climatic effects resulting from this huge increase in carbon dioxide is suggested by mathematical experiments with atmospheric general circulation models. For a doubling of carbon dioxide concentration, the initial results obtained by means of a still far from realistic model indicate a global temperature increase of more than 2°C, with larger changes in the polar regions. Simple extrapolation would suggest, therefore, that the yet higher concentrations projected in the next century could produce temperature increases comparable with the transition from the last ice age to the present (about 6°C in global mean temperature). Comparably large changes in the distributions of ice and snow, boundaries of agricultural areas, sea levels, and other phenomena could be expected. These projections, made with existing knowledge and inadequate present-day models based on that knowledge, *must be used with great caution.* Many important factors, processes, and interactions are omitted or crudely approximated in the models. For example, coupling between temperature changes and cloudiness could greatly affect the magnitude of temperature changes, but it still is not known whether they would be positive or negative. Experiments with present models should, therefore, be considered as warning flags not as authoritative prescriptions.

Nevertheless, it appears highly unlikely that man can inject such

*C. D. Keeling and R. B. Bacastow (1977). Impact of Industrial Gases on Climate, Geophysics Research Board, in *Energy and Climate*, National Academy of Sciences, Washington, D.C.

large amounts of such an important constituent as carbon dioxide into the atmosphere without inducing *some* sort of environmental change. Moreover, as suggested above, unexpected climatic alterations will greatly stress future societies. It is, therefore, possible that adverse climatic changes may be the primary limiting factor in the use of fossil fuels for energy. Such a limit may well force either drastic alterations in human lifestyle or an earlier transition to alternate energy sources than would be dictated by economic considerations alone.

For these reasons, an understanding of climate and the mechanisms of climate change is crucially important to the design of future world societies. Through scientific efforts of the last few decades, a sound foundation for research has been built, and there are encouraging indications that the principal outstanding problems are tractable. Recent national and international efforts have produced rational and feasible proposals for monitoring the climate for fundamental research in climate changes and for modeling. These plans must be implemented if we are to build a sound basis for satisfying mankind's future needs for energy.

VIOLENT WEATHER*

Although the weather influences human activities in many important ways, it may often go unnoticed. A summer season with temperatures and rainfall generally below a statistical "normal" may get little overt attention, but the effects can be significant in terms of human comfort, power consumption for air conditioning, and crop yields. On the other hand, death-dealing violent storms such as tornadoes and hurricanes make the headlines from coast to coast.

In an average year, the number of deaths and the amount of property damage caused by severe storms are disturbingly high (see Table 1). During some years, the costs are very much greater than the averages. For example, in 1969, Hurricane Camille hit Mississippi and Louisiana and caused over 250 fatalities and more than a billion dollars in property damage. More recently, Hurricane Agnes in 1972 did more than $3 billion in damages. During April 3–4, 1974, some 148 tornadoes occurred over a large part of the eastern United States, leaving 307 people dead and destroying more than half a billion dollars of property. Many other examples could be given to illustrate the devastating effects of violent weather.

*Major contributions to this section by Edwin Kessler, National Severe Storms Laboratory, National Oceanic and Atmospheric Administration.

TABLE 1 Storm Deaths and Damage in the United States[a]

Type of Storm	Recent Average Annual Deaths	Recent Average Annual Property Damage (1975 Dollars)
Tornado	100	150 Million
Lightning	150	200 Million
Hail	—	500 Million[b]
Hurricane	75	500 Million

[a]Source of data: Edwin Kessler, Director of the National Severe Storms Laboratory, NOAA.
[b]About 1 percent of agricultural production in the United States.

Since the turn of the century, even though the population has more than tripled, there has been a downward trend in the number of storm-caused deaths in the United States. This encouraging result can be attributed to improvements in techniques for observing and predicting storms, important advances in procedures for broadcasting storm warnings, and public education programs for advising the public on appropriate safety precautions.

The decrease of casualties, not surprisingly, has been accompanied by a rise in property damage. This reflects an increase in the number and the spatial density of buildings and farms vulnerable to strong winds, hail, and floods.

In order to reduce the costs inflicted by violent weather, it is necessary to learn more about the storms and how to deal with them. In more specific terms, we need to address the following goals:

1. The development of improved techniques for observing the formation, evolution, and movement of severe storms.
2. The development of improved techniques for predicting the place and time of occurrence of storms. This would include the use of mathematical models based on realistic theories.
3. The development of improved techniques for accurate and timely warnings to those people threatened by violent weather.
4. Education of the public on appropriate responses in order to reduce vulnerability to injury and loss of property and the establishment of building standards for houses and mobile home parks to make them more resistant to violent weather.
5. The development of techniques to reduce the hazards of severe storms by means of cloud seeding or other weather-modification procedures.

Substantial progress has been made in some of these areas over the last few decades. This accounts for the reduction of casualties in the United States in the face of increasing population.

Goals 1, 3, and 4 are mostly engineering or educational in nature and can be achieved given sufficient money and time. The forecasting problem (goal 2), however, is a particularly difficult one because it requires significant advances in our comprehension of the physical processes in severe thunderstorms, tornadoes, and hurricanes. Goal 5, dealing with the modification of violent storms, is the most difficult of all. We still are not sure that major storm systems can be modified. Nevertheless, it seems reasonable to expect that once we understand storms well enough to better predict their occurrence, we should be able to introduce some types of perturbations to modify their behavior. As noted in Chapter 4 in the section on weather modification, there is evidence indicating that thunderstorms can be influenced in important ways and that hurricanes can be weakened. Unfortunately, the evidence is still not sufficient to serve as the basis for a program in severe-storm engineering. The potential value of a reliable technology for mitigating the hazards of violent weather is so great that an aggressive, well-funded national research commitment is justified.

Violent Thunderstorms

A common feature of most violent storms is that they form in a moist, unstable atmosphere. On such occasions, the air temperature is relatively warm near the ground and decreases rapidly with height. When the atmosphere is unstable and air is caused to rise as a result of ascent over a hill or mountain, or a perturbation in the wind field, the air is accelerated upward. If it is moist, little ascent is needed before cloud droplets are formed by condensation of water vapor on tiny nuclei in the atmosphere. When condensation takes place, heat of condensation is released into the atmosphere. This is called "latent heat" and is the major source of energy for the severe air motions found in violent storms.

When the atmosphere is sufficiently moist and unstable, updrafts can reach into the lower stratosphere at altitudes exceeding 15 km. In such circumstances, the results may be towering thunderstorms producing copious rain, lightning, and thunder; sometimes hail occurs as well.

On some very unstable days, particularly over the central United States, the wind speed increases with height and turns from mostly southerly near the ground to mostly westerly aloft at altitudes exceeding 8 to 10 km. When this occurs, thunderstorms can be unusually large, oriented in the form of squall lines, and may last for many hours. Warm, moist air moves into the storm at low levels and passes upward through the thunderstorms and out of them at high levels. These are known as supercell storms, which can produce torrential rain, large hail, and, in some cases, tornadoes.

Lightning

As already noted, there are various types of thunderstorms, which differ in size, duration, internal structure, and the violence of the weather they produce. Every thunderstorm, by definition, exhibits at least one lightning stroke and the associated thunder. There is a great range in the degree of electrical activity. In some storms, particularly those over the mountains of the western United States, there is little rain reaching the ground but fairly extensive cloud-to-ground lightning. As a result, such storms frequently start forest fires in dry timber and grasslands.

Lightning is essentially a massive discharge of electricity—a giant spark or electric arc. It serves to neutralize an accumulation of electric charge in the clouds. Although Benjamin Franklin, some 200 years ago, demonstrated the electrical character of thunderstorms, we still are not sure of the processes by which the electric charge is separated. It has long been known that the upper parts of thunderstorms are mostly positive, while the central and lower parts are negatively charged. In the rain near the cloud base, there is sometimes a secondary center of positive charge.

Various theories have been proposed to account for thunderstorm electricity. Most experts agree that the interaction of ice particles with other ice particles and supercooled water drops* leads to the separation of negative and positive charge, with the latter being carried to upper levels by the updrafts. When the quantities of charge are sufficiently great, the electric potential differences can exceed the so-called "breakdown potential" and a discharge is initiated.

Over the last two decades, as a result of new observational techniques, a great deal has been learned about the nature of lightning itself. For example, we know that the lightning channel is only a few centimeters in diameter and that its peak temperature is about 25,000°C. Some flashes are composed of only a single stroke and last for only a few hundredths of a second, while others are composed of as

*Those composed of water but having temperatures below the nominal freezing point at 0°C.

many as 40 strokes and last for several seconds. In some instances the currents continue to flow for relatively long periods. These are the flashes most likely to produce fires, because they generate a great deal of heat as they pass through trees or buildings.

Notwithstanding the impressive advances in understanding lightning and thunder, much more remains to be learned. Only by means of comprehensive observation measurements of thunderstorms can uncertainties about mechanisms of charge separation in thunderstorms be resolved. It is essential to develop a better understanding of the mechanism by which lightning discharges are initiated and maintained and the role of lightning, if any, in precipitation formation processes.

Hailstorms

Over the central and eastern parts of the United States, thunderstorms commonly produce heavy rain. In most cases, the storms are small and fairly isolated with typical lifetimes of an hour or less. On less frequent occasions, there are intense long-lasting storms, either as fairly isolated systems or in the form of long lines. They can maintain themselves for many hours as they sweep across the country and sometimes produce hail and tornadoes.

As noted earlier, the most prolific hailstorms are called supercell storms. It has been found that low-level air moves into them from the leading edge up through the storms and mostly out from the top. The storms can last for a long time, with a slowly varying configuration of upwards air motion. The strong, persistent updrafts allow large hail to grow. Over the Great Plains, where hail is frequent, most of the crop damage is caused by just a few supercelled hailstorms each year.

Research in various parts of the world, for example, in the United States over eastern Colorado, southwestern Nebraska, and southeastern Wyoming, is yielding information about the detailed properties of hailstorms and hailstones. Scientists of the National Hail Research Experiment, managed by the National Center for Atmospheric Research (NCAR) in Boulder, Colorado, and their university associates are collecting unique observations by means of modern radar, instrumented airplanes, and a network of ground stations. A specially reinforced instrumented airplane is being used by South Dakota scientists to fly through hailstorms and measure their internal properties.

Laboratory analyses are being made of the structure of hailstorms to ascertain the properties of the ice embryos on which stones form. Studies are in progress of the characteristics of the ice composing the alternate layers of clear and milky ice that characterize many hailstones.

Theoreticians are working on mathematical models of hailstorms. They seek to relate the information and growth of hailstones to the many interacting processes within a storm.

Most hailstorms yield small ice particles less than the size of sweet peas, and sometimes the hailstones are the size of oranges or even greater. Accounting for such a range of hailstone sizes is a formidable challenge. Learning to predict the occurrence of the storms and ultimately to control them are even greater challenges.

Tornadoes

Tornadoes are probably the most frightening storms of all. Although the data show that on the average they kill fewer people and do less damage than lightning, hail, or hurricanes, the nature of tornadoes creates fear. They strike suddenly, usually with little warning, and in a few minutes destroy virtually everything in the path of the funnel.

Record-sized hailstone collected during storm at Coffeyville, Kansas, in September 1970. (Courtesy, NCAR)

Tornado funnels are small, are generally on the ground for only a few minutes, and the maximum winds are about 200 mph (90 m/s), with gusts to as much as 100 mph (45 m/s) higher. Over the last 10 years there has been an average of about 700 tornadoes reported per year. As you would expect, there are ranges of sizes and durations. Most of the fatalities and damages are caused by a small number of large, long-lived storms. According to two American experts, Allen Pearson of NOAA and T. T. Fujita of the University of Chicago, during the decade ending in 1970, the largest tornadoes account for 2 percent of all tornadoes and for 85 percent of the fatalities. Pearson refers to the strongest 5 to 10 percent of the storms in the United States (about 50 per year) as "maxi-tornadoes."

Until the late 1940's, the U.S. Weather Bureau did not issue tornado forecasts. It was recognized that they were difficult to predict, and there was concern that the forecasts would frighten the public, often unnecessarily. Over the succeeding years, there have been major improvements in methods for forecasting storms, detecting them, and communicating tornado watches and warnings to the public.

Tornado forecasters are confronted with several major problems. An inadequate understanding of the nature of tornadoes and the processes of development and dissipation certainly contribute to the difficulties.

An approaching tornado moving toward the left. (Courtesy, NOAA)

Even if we knew much more about severe thunderstorms and the sequence of events leading to tornado occurrence, the spacing of observations in distance and time would put limits on tornado forecasting. The balloonborne rawinsondes (which measure temperature, humidity, and wind structure of the atmosphere) are launched from stations separated by several hundred kilometers and usually take observations only twice a day. It is not surprising that these observations lead to forecasts that can do no better than call for tornadoes to occur over areas of perhaps 65,000 square kilometers (about 25,000 square miles).

Rawinsonde observations at shorter time intervals, and at more stations, should help to reduce the size of the predicted area of violent weather.

Notwithstanding their limitations, the existing forecasts are of great value because they focus attention on those regions of greatest vulnerability. To locate tornadoes more precisely, it is necessary to use visual observations and remote detection devices such as radar, weather satellites, or radio receivers, which detect the radio signals emitted by the storms.

Observation of Tornadoes

It is unfortunately true that even in these years of space exploration, most tornadoes are detected visually by ordinary citizens who spot the funnels and notify local officials. On some occasions, those who observe the storms do not have a chance to notify anyone because they are trying to save themselves and their families. At other times, telephone lines are down and communication is impossible. It should not be necessary to rely on visual observations to verify the occurrence and the location of storms as devastating as tornadoes.

T. T. Fujita, at the University of Chicago, has proposed that observations of the behavior of thunderstorms from satellites can indicate regions where violent weather can be expected. We can anticipate that as the spatial and time resolution of satellite cloud observations improve, and more detailed measurements become available, they will play an important role in the location and monitoring of regions of hazardous weather. In the years to come, specialized radar carried on satellites should play a part in the identification of violent storms.

It is well known that electrical discharges, particularly in the form of lightning, cause the emission of radio waves, which take the form of static on radio receivers. They used to be called "atmospherics,"

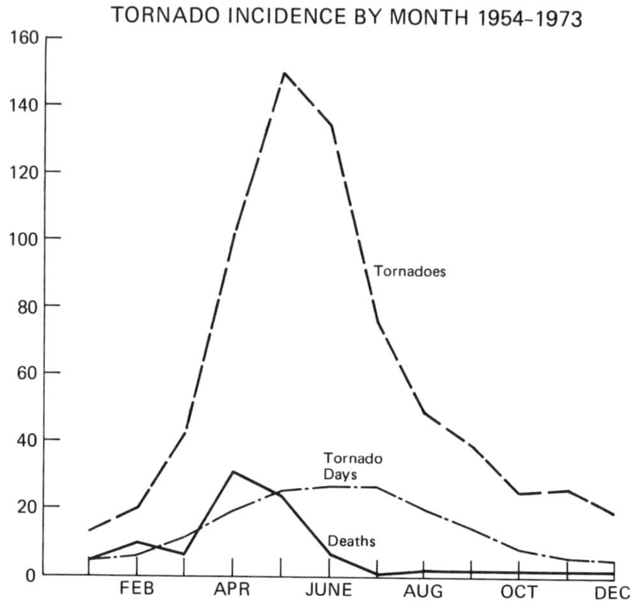

Average monthly incidence of tornadoes, tornado days, and loss of life in the United States. (Courtesy, NOAA)

which over the years has been contracted and transformed to become "sferics."

In about 1951, Herbert L. Jones, at Oklahoma State University, proposed that the sferics from tornado storms were identifiably different from those produced by thunderstorms, even intense ones. Various groups have investigated the feasibility of employing sferics signals to discriminate between tornadoes and thunderstorms.

At this time, it can be said that although appropriate sferics receivers can identify some tornadoes, they cannot detect all of them. Some tornadoes, especially the small, short-lived ones, do not have electrical properties distinctly different from nontornadic thunderstorms.

Since World War II, radar has been used to observe the location, size, movement, and intensity of thunderstorms. It has been employed extensively in research and day-to-day weather operations. A network of radars throughout the United States monitors the formation and behavior of thunderstorms and other storms.

A conventional radar of the type used by the National Weather Service detects the presence of water drops and ice particles. In general, the greater the size and concentration of the precipitation particles the more intense the radar echo. It has been found that long-lasting thunderstorms extending to great altitudes and producing intense echoes are likely to have associated violent weather—hail and tornadoes. When a weather forecaster observes such thunderstorm echoes moving across a region within which tornadoes have been predicted, he can localize the areas where tornadoes are most likely to occur.

Since a conventional radar cannot specifically identify a tornado, it is necessary to use characteristics of the echoes to infer the presence of a tornado. When a very intense echo is observed in an area where tornadoes are expected, it may be inferred that a violent storm is present or is shortly to form. If at all possible, attempts are made via radio or telephone to obtain visual observations from state police or other individuals in the vicinity.

Occasionally the shape of a radar echo is a good indication of the presence of, or the impending formation of, tornadoes. The most reliable such indicator is a hook-shaped echo extending from a large thunderstorm echo. Unfortunately, such distinctive indicators do not occur often enough. Most of the time, the shape of the echo is of little value in identifying tornadoes.

One of the greatest values of radar is to track tornado-producing thunderstorms. As noted earlier, some traveling storms produce a series of funnels. Once the "mother storm" is identified, it can be followed and people in its path given warnings of its approach, perhaps an hour or two in advance of the arrival of the storm.

A Doppler radar is one that can measure the same quantities as can a conventional radar, but in addition it can measure the speed of the targets (e.g., raindrops) toward or away from the radar. Since one of the most characteristic features of a tornado is the high wind speeds, it seems reasonable to expect that a Doppler radar would be suited ideally for tornado detection. Such an idea has been discussed for more than two decades, but only in recent years has it been vigorously pursued.

Before examining some promising advances, it is in order to note some limiting characteristics of Doppler radar. One problem has to do with the region of coverage. A great many radar sets would be needed to keep under surveillance the entire tornado-prone regions of the United States. On the other hand, the population centers, and therefore most of the people, could be covered with a reasonable number of radars.

A factor that has been holding back the development of a Doppler radar for operational tornado detection has been the difficulties involved in the processing of the radar data and the display of the relevant information quickly and in a fashion that is easy to interpret. A weather forecaster wants to know, as soon as possible, the precise location of high-velocity vortices that might be tornado funnels. Fortunately, in the last few years, largely as a result of research by Roger Lhermitte, now at the University of Miami, and scientists at the Air Force Cambridge Research Laboratories and at the National Severe Storms Laboratory, considerable progress has been made in this area of investigation.

If two or more Doppler radars are employed to observe the same storm, it becomes possible to obtain three-dimensional air-flow patterns. Clearly, there are problems in adapting multiradar methods for operational use, but they are not insurmountable and need to be pursued.

Recent research clearly indicates that an appropriately designed Doppler radar would be of great value in the location of tornadoes in the atmosphere. In some cases, it may be possible to observe the early stages of tornado development as much as 20 minutes before the funnel reaches the ground. Such information would be of great value to tornado forecasters.

Dissemination of Warnings

Techniques for the accurate location of severe local storms are of little value unless there are effective procedures for disseminating warnings to the people being threatened by the storm. The present "tornado watch," a forecast issued some hours before storms are expected, may cover an area of about 65,000 square kilometers (25,000 square miles), which encompasses large numbers of people, only a small fraction of whom are likely to be affected even if tornadoes do occur. Many people are kept in a state of anxiety and fear. In some cases, too many "watch" forecasts, not followed by tornadoes, can lead to complacence.

When a tornado is observed and reported to the National Weather Service, a "tornado warning" is issued. In this case, the time available for broadcasting effective warnings is short, and reaction times by citizens must be fast.

A Panel on Severe Storms of the National Research Council is currently considering the wisdom of having a "tornado alert" issued at an intermediate time between the "watch" and the "warning." It would serve to refine and make more precise the earlier message and would give more notice than the warning message.

Specific techniques for disseminating forecasts need to be reexamined in terms of their effectiveness in actual practice. Most reliability is placed on radio and television, but power outages often render these media ineffective.

Tornado Research

In summary, it seems clear that in order to mitigate the hazards of severe thunderstorms, particularly tornadoes, it is necessary to develop and use more effective remote-sensing techniques. It is particularly important to pursue aggressively the following observation techniques and communications problems:

1. It is necessary to design, construct, and test a Doppler radar for the detection of tornadoes. The goal should be a radar that can be used operationally by the National Weather Service.
2. There should be expanded research on the use of multi-Doppler radar systems for the observation of severe thunderstorms and the hazardous weather they produce.
3. There should be an enlarged program of research on the utilization of weather satellites for the detection of violent weather and as a communication link for broadcasting severe-storm forecasts.
4. Communication experts, working with meteorologists, should seek more effective procedures than those now in use for disseminating tornado forecasts to the public.

Hurricanes

Whereas tornadoes are small, short-lived storms forming over land, hurricanes are large and long-lasting and develop over oceans. They occur most frequently in late summer and early autumn over warm tropical oceans, both north and south of the equator. The names "typhoons" and "cyclones" are the names given to hurricanes over the eastern Pacific Ocean and over the Indian Ocean, respectively. Many tropical storms and weak hurricanes occur off the west coast of Mexico. Occasionally they move northward into southern Arizona. Although the overland trajectory weakens the storms, they still can yield widespread, heavy rain, which is of great value in maintaining the ecology of the Sonoran desert.

Atlantic hurricanes come into being over the warm water of the tropical ocean. Recent analyses, made possible by satellite observations, indicate that many tropical storms may be initiated by weather

Aerial photographs of Xenia, Ohio, high school before and after tornado of April 3, 1974. (Courtesy, NOAA)

disturbances moving westward over the southern part of the North Atlantic. When atmospheric conditions are appropriate, counter-clockwise air circulations are started. As the storm enlarges and the winds become stronger, spiral bands of rainclouds are formed around a central region of winds moving around a relatively calm, cloud-free area called the "eye." It might be 20 to 50 km in diameter.

Hurricanes average about 200 km in diameter at maturity and last for a period of one to two weeks. Typically they move toward the west as they develop in the belt of prevailing easterly winds. The storms then follow curved paths toward the north and northeast. Most of them do not hit North America but instead spend their entire lifetimes over the Atlantic. As they move over cooler water, the supply of energy is diminished and the hurricanes become weaker. Similarly, storms moving over land become separated from the warm ocean water—the chief energy source—and dissipate rapidly.

Although it is well known that hurricanes are very destructive, it

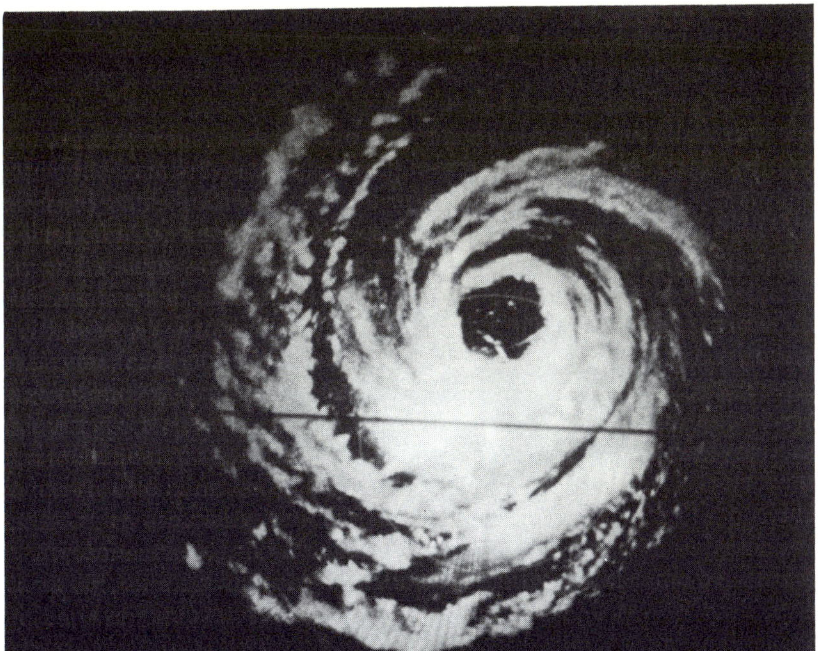

Radar image of hurricane showing converging circular bands of clouds and the "eye" of the storm. (Courtesy, NOAA)

should be recognized that they also do a great deal of good by supplying much needed rain to many parts of the world.

The destructive features of hurricanes are the strong winds, the surge of ocean water over low-lying coastlines they produce, the intense rainfall, and the tornadoes that often form within the hurricane circulation.

Of these various factors, the most devastating is the storm surge. As the storm approaches land there is a steep rise of ocean level, which is called the "storm surge." It is in the form of a wave more than 3 m high. The surge of ocean water over low-lying coastlines leads to the drowning of any living thing not evacuated to higher ground. A hurricane that hit Bangladesh in November 1970 took the lives of over a quarter of a million people who were trapped by the rapidly rising water.

When a hurricane has moved northward and its winds have diminished, it is called a tropical storm. Winds cease to be a major hazard, but rainfall still is a major threat, particularly when the storm passes over mountainous terrain.

In June 1972, the remnants of Hurricane Agnes caused torrential rains and massive flooding in Virginia, Maryland, Delaware, Pennsylvania, New York, and West Virginia. In some places, more than 14 inches (356 mm) of rain fell in 24 hours. Over the two-week period of its existence, Agnes, which started in the Caribbean Sea, caused 118 deaths and more than $3 billion in damage, primarily as a result of the flooding.

Predictions of the tracks of hurricanes are of enormous value in giving people in vulnerable areas time to move to higher ground. The forecasting techniques take into account the statistical behavior of past hurricanes and calculations of the movement and intensity of storms by means of mathematical models of the atmosphere.

The combination of statistical with dynamical techniques provides forecasts couched in probabilistic terms, with ellipses to define the area where the storm is expected to be at some future time. As forecasts improve, the ellipses become smaller. At present, forecasts made 12 hours in advance present an average error in landfall of about 75 km.

From the time a tropical storm or hurricane is first identified, advisories are issued at 6-hour intervals. When a hurricane's path is projected within 36 hours to lie over an inhabited area, the advisory includes a hurricane "watch," a statement that specifically alerts the potentially affected population but stops short of calling for protective measures. Because of considerable uncertainty in path projections so far in advance, the watch region is usually much larger than that

Major flooding at Wilkes-Barre, Pennsylvania, following passage of Hurricane Agnes, June 1972. (Courtesy, American Red Cross)

actually affected by the storm at a later time. When the storm is within 18 hours of projected landfall, the National Weather Service issues "warnings." They call for action and may suggest securing loose property and evacuating low-lying areas before rising water closes roads. The absence of an adequate warning system and the subsequent failure to evacuate in time cost 6000 lives in 1900 when a hurricane storm surge hit Galveston, Texas.

Such a disaster should never happen again in the United States. Modern techniques of remote sensing from satellites make it possible to detect and follow hurricane development anywhere on earth. It still is not possible to make accurate wind measurements within the hurricane from satellite data; this problem is being studied. In the meantime, long-range radar from ground stations observe hurricane position, sizes, and, in some cases, intensity, when the storm center is within a few hundred kilometers of the radar site. Specially instrumented airplanes are flown into hurricanes when they are within range of airports in order to make measurements of wind speeds and other storm properties.

It is accurate to say that the hurricane observation, forecast, and warning service in the United States is an excellent one. At the same time, meteorologists in the hurricane business recognize that it could be improved if we knew more about the factors governing the initiation, growth, and movement of hurricanes. An improved warning service will require an increase in the accuracy of predictions of the place and time when hurricanes will pass onto the land.

Another necessary task is to develop improved procedures for forecasting the behavior of hurricanes as they pass inland over hilly or mountainous terrain. It is particularly important to learn to predict the rainfall they are likely to produce in order to protect the public from unexpected floods.

As is noted in Chapter 4 in the section on weather modification, research is in progress to develop techniques for decreasing the intensity of hurricanes by means of cloud seeding. The goal of this research is to reduce peak wind speeds and the storm surge and consequent coastal flooding by seawater. If this could be done, property damage and casualties could be significantly reduced. It is hoped that storm damage can be diminished without causing reductions in beneficial rainfall.

In some parts of the world, notably southeast Asia, hurricanes still represent a major threat to life and property. Whatever we know and learn about the formation, growth, and movement of Atlantic hurricanes or Pacific typhoons will be of great value wherever these storms occur.

3

Needs for Skillful Predictions of Weather and Climate

WEATHER PREDICTION

Perhaps no other aspect of the atmospheric sciences has a greater impact on the public consciousness than the weather forecast. Indeed, because of widespread public interest in tomorrow's weather, most news broadcasts would be considered incomplete without a few words about the weather. Of course, some of the public interest in weather predictions represents a general curiosity about the future. More importantly, interest in weather forecasts stems from the many ways in which human activity depends on or is affected by the weather. In the case of violent storms, such as tornadoes and hurricanes, the forecast can mean the difference between life and death.

Weather prediction is not an exact science, as evidenced by the many forecasts that prove to be inaccurate. There are many scientific, technological, and related problems that complicate the prediction problem. In this discussion, we shall be concerned not only with the weather prediction itself but also with the nature and magnitude of the difficulties still to be overcome. This section will also include assessments of current predictive skills and the economic and social values of weather forecasts and warnings.

Prepared by Jack C. Thompson, San Jose State University.

The Basis of Weather Prediction

Weather prediction is among the oldest of scientific problems. Early forecasting methods were primarily empirical: a repeated sequence of observed events, each followed by a certain kind of weather, was assumed (often erroneously) to prescribe a cause and effect relationship. By the latter part of the nineteenth century, it was recognized that in principle the problem might be attacked by making use of basic physical laws, just as in the prediction of the orbits of the planets. By the beginning of the twentieth century, scientists had discovered the basic physical principles upon which today's forecasting is based. These principles apply equally well to all motions in the atmosphere, from gentle leaf-rustling breezes to intense tornadoes or hurricanes. However, these laws of motion achieve their generality by describing explicitly only the most minute changes that take place from one instant to the very next for each infinitesimal element of air. To understand the pattern and evolution of the motion of any sizable portion of the atmosphere, it is necessary to piece together, or integrate, the explicit descriptions of small local changes of the various parts to deduce the architecture of the motion of the whole. This demands a mathematical procedure of enormous difficulty to move from the principles to their useful implications.

It was not, in fact, until the development of the electronic computer in the 1940's that the first successful albeit simple and imprecise, numerical solutions were attained. Initial progress was relatively rapid, and the first operational use of electronic computers for producing weather predictions occurred in 1955.

Progress in the development of modern meteorological prediction over the last two decades has resulted from a gradual increase in the sophistication and realism of the numerical models. Each painstaking modification has brought us a step closer to describing the characteristics of the real atmosphere. The present models require several orders of magnitude more computations than the earliest ones; this increase in complexity would not have been possible were it not for concomitant developments in computer technology.

Notwithstanding major advances, difficulties still remain. One of the most serious difficulties is the lack of observing stations over the oceans, particularly in the tropics and in the southern hemisphere. Remote temperature soundings from satellites have done much to alleviate this problem in middle and high latitudes, although there still remain some unsolved problems in meshing the satellite soundings together with the soundings derived from the conventional radiosonde

A model of the X4, Advanced Scientific Computer at the Geophysical Fluid Dynamics Laboratory (NOAA), Princeton University. (Courtesy, NOAA)

network. Over the tropics, where the temperature variability is too small to be resolved by remote sensing from satellites, cloud motions derived from sequences of satellite images 30 minutes apart are providing a great deal of useful information on the wind field. As noted earlier in this report, during 1978–1979 a major international effort is expected to upgrade the global observing system.

Another important source of error is the limited spatial resolution of the numerical models. The current models are incapable of resolving weather systems with horizontal scales smaller than a few hundred kilometers. Unfortunately, the unresolved, small-scale features include weather phenomena of considerable importance for forecasting, such as severe thunderstorms and precipitation patterns in mountainous regions. The problem is compounded by the fact that certain physical processes like cumulus convection, which take place on a rather small scale, ultimately influence the time evolution of the large-scale wind field. Thus, the lack of spatial resolution in the models contributes to errors in the forecasts of the large-scale components of weather patterns.

In order to cope with this resolution problem, three different approaches are being taken:

1. Higher-resolution models are being developed, including models with "nested grids," which can provide extra fine resolution over

limited regions of special interest. The price of higher resolution is an increase in the amount of computation; doubling the horizontal resolution requires eight times as many computations. With modern computers approaching fundamental speed limitations imposed by the speed of light, it is clear that horizontal resolution cannot be increased indefinitely. Nevertheless, it is likely that the next generation of "fine-mesh" models now under development will represent a significant advance over the present models.

2. Statistical methods are being used to relate the probability of occurrence of important small-scale weather events, such as severe thunderstorms, to the large-scale weather patterns forecast by the models. These so-called "Model Output Statistics" can also be used to provide forecasts of rain and snow amounts, wind speeds, pollution levels, high and low temperatures, and other weather phenomena in terms of probability information tailored to meet the needs of various users.

3. Researchers are seeking a greater understanding of small-scale atmospheric processes, such as cumulus convection and boundary-layer turbulence, with hopes of finding ways of representing or parameterizing these processes in the models in terms of the large-scale weather patterns. In this manner, their effects on the time evolution of the large-scale patterns can be taken into account without resolving them explicitly in the models. The GARP Atlantic Tropical Experiment (GATE), conducted off the coast of Senegal during the summer of 1974, was motivated by the need for developing a scheme for parameterizing deep cumulus convection in the tropics. During GATE, instrumented aircraft revealed the internal structure of deep convective clouds, while satellites and ships monitored the large-scale environment. Data from the experiment are now being processed and analyzed. Over the course of the next few years, they will be used to test various parameterization schemes that have been proposed for representing the influence of the convection upon the large-scale wind field.

Even if the problems of inadequate initial data and limited spatial resolution could be overcome (which, of course, they never will be completely), there would still be a limit to the range of accurate, day-to-day weather forecasts. This so-called limit of deterministic predictability is fundamental to the nature of atmospheric motions. Some rather simple mechanical systems are highly predictable, over extended time periods; for example, the motions of the planets can be predicted far into the future. In this respect, atmospheric motions are more analogous to the motions of a marble in a pinball machine. Two

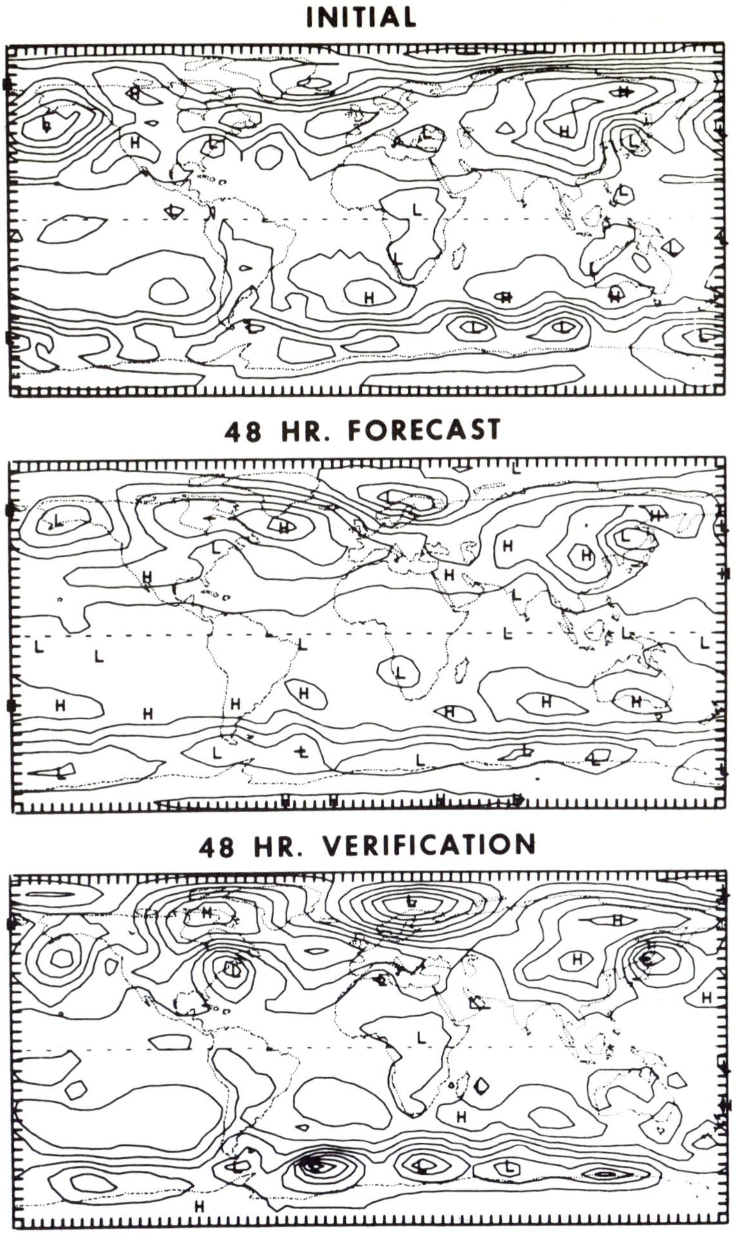

Results of a numerical experiment in a 48-hour prediction of sea level pressure with the global circulation model of the National Center for Atmospheric Research. (Courtesy, NCAR)

marbles starting with practically the same initial velocity may start along almost the same path, but with each passive obstacle that they encounter, their paths become more dissimilar until eventually all resemblance is lost. Similarly, in a forecast model, even the smallest uncertainty in the initial state of the atmosphere would gradually grow larger as the forecast interval was extended. Eventually, it would grow so large as to render deterministic forecasts worthless. Current estimates of the predictability time of the atmosphere range from about 5 to 10 days.

For periods beyond a few weeks, it appears likely that the best we can hope for in the way of forecasts is statistical information such as average temperature and precipitation over rather large areas and extended time intervals such as a week, a month, or a season. Limited though such information might be, it would still be of great economic value to agriculture and other industries that are strongly weather-dependent.

The Skill of Present Weather Forecasts

At this point, it is appropriate to ask what skill has been attained by the meteorological profession in predicting the weather. In order to answer this question, the term "skill" needs to be defined. For forecasts up to about a day in advance, *persistence* serves as a useful "control" against which to measure the accuracy of weather forecasts. Persistence is simply a forecast that the present weather will continue. Thus if a forecasting technique is useful for predicting weather *changes*, it must exhibit skill relative to persistence. For longer-term forecasts, it is more appropriate to measure skill relative to a "control" based on a linear combination of persistence and climatology, where the latter represents a forecast that the future weather will be "normal" as defined by past climatological statistics. The longer the forecasting interval, the less the weight that should be given to persistence and the more to climatology in the "control" against which skill is measured. If a long-range forecast is superior to climatology, it is said to have skill.

For periods up to two days in advance, weather predictions have sufficient skill to serve as a basis for a wide range of operational decisions by the forecast user. Such forecasts are of greatest usefulness when accompanied by estimates of the probability that an operationally critical weather event will occur. The general geographical area and time period of small-scale, severe weather phenomena can be predicted up to 24 hours in advance. However, the precise location or time of occurrence of such phenomena cannot be forecast with any degree of skill beyond one to two hours. Similarly, accurate forecasts of infrequent weather, such as heavy snow, sleet, or damaging winds are usually limited to periods not exceeding 24 hours.

For periods up to five days, daily temperature forecasts of moderate skill and usefulness can be made. Precipitation forecasts have an equivalent level of skill up to three days, but the skill drops to marginal levels on the fourth or fifth days.

For periods of more than five days, average temperatures for periods of a week up to a month or season can be predicted with slight skill, and there is some skill in the prediction of total precipitation amounts for periods up to a week in advance. However, predictive skill for periods beyond these times is either marginal or has not been demonstrated. It should be noted that certain meteorologists have argued that the skill in some of these longer-range, average predictions may be mostly a result of an ability to skillfully predict the weather over the first five days or so.

The skill of weather forecasts has shown a modest but significant improvement over the past 20 years. Improvements have been most notable in the accuracy of warnings concerning the most destructive storms: hurricanes and tornadoes. Although damage to property, which is almost unprotectable, has increased materially during the past 50 years, the number of lives lost from such storms has greatly decreased. Furthermore, this saving of life has been achieved despite a large population increase in the storm-affected areas.

The Value of Weather Prediction

During the past four decades, meteorological technology has become increasingly complex and sophisticated. Nonetheless, annual *per capita* cost for the country's meteorological activities and services is only about $2.50. The annual budget for meteorological services and supporting research in the United States at present is close to $700 million. For the world as a whole, it is the equivalent of about $2 billion.

Although part of this expenditure provides support for basic research and related activities, nearly 80 percent is associated directly or indirectly with weather forecasting, the applied branch of the science. These weather predictions serve many purposes: they help communities to protect their citizens against disastrous storms; they allow aircraft pilots to utilize favorable jet stream winds; they enable ocean-going vessels to avoid dangerous seas. Weather information allows construction companies to specify labor needs for outdoor exposed

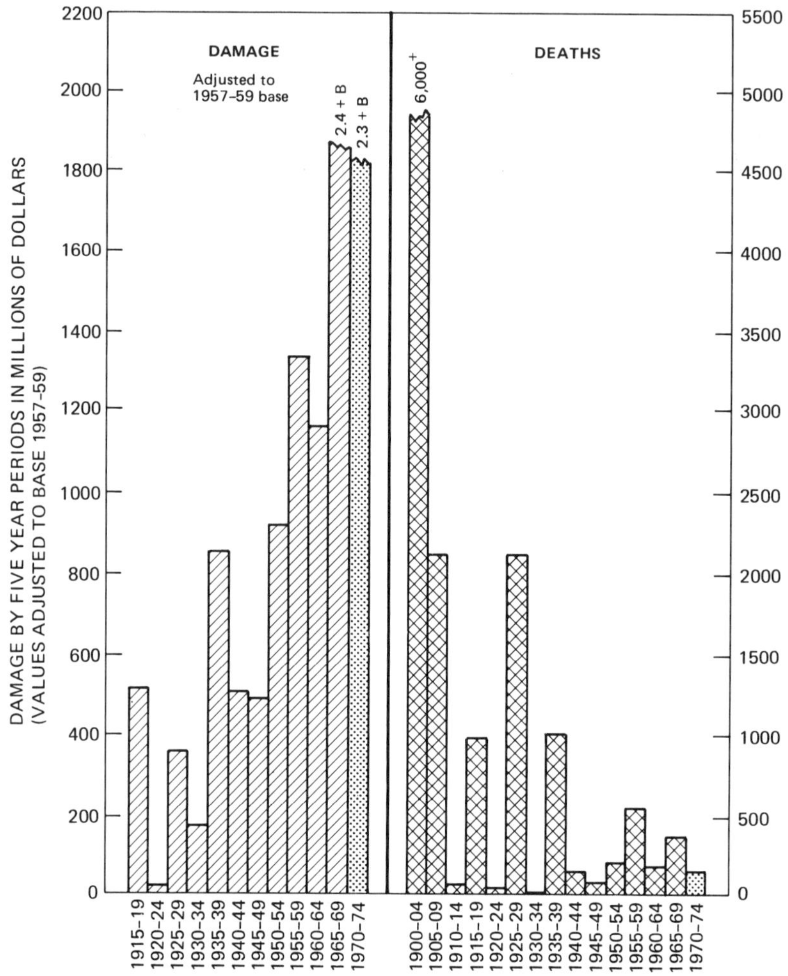

Damage and loss of life trends from hurricanes in the United States, showing the significant reductions in death through improved warning services. The growing vulnerability of certain regions of the country is revealed by the marked increases in property losses (updated and adapted from R. C. Gentry, 1970: Hurricane Debbie modification experiments. *Science 168*, pp. 473–475).

activities, and farmers are aided in determining whether certain activities can be effectively undertaken. Forecasts even help the family to decide whether to proceed with a planned vacation or outing.

Are the weather forecasts worth their current costs? Will prospective advances in the science improve the usefulness of weather predictions? What would be the potential value of improved weather predictions?

In attempting to answer these questions, a number of relevant factors must be recognized. First, nature's destructive forces are occasionally so great that it may be impractical, or even impossible, to prevent property damage. This is exemplified by tornadoes, which, in the United States, cause an annual average of $150 million in virtually unavoidable property destruction. Another problem is associated with certain inadequacies in the response of the public to warnings.

Considerable thought has been given in recent years to methods of assessing the value of weather services. In 1968, the World Meteorological Organization, through its member nations, provided estimates of the current value of meteorological services to their countries. In Australia, for example, with an annual budget of $11 million (Australian) for weather services, the savings associated with weather forecasts and warnings were estimated to be $300 million. In France, the annual budget of 100 million (old) francs produced savings of 2000 million francs. A benefit of 20 times the cost of meteorological services was estimated for the United Kingdom, and improved forecasts associated with planned new computers were expected to increase this figure. In the Soviet Union, a two-year program involving the use of weather information for routing ocean shipping produced a savings of 4 million rubles, while the cost of the service was about 320,000 rubles. The Federal Republic of Germany reported a 3 percent reduction in the cost of long-distance flights as a result of efficient use of upper-level wind forecasts. In the developing countries, meteorological information was felt to be especially beneficial, since only a small investment would be required to compile climatic statistics the use of which could bring about large returns in the proper planting of crops.

From individual estimates and fragmentary data such as these, it seems evident that the benefit-to-cost ratio for *present* meteorological services is quite large. Indeed, such services are among the most efficient applications of today's science and technology. However, recognizing that improved services, such as more accurate forecasts, may be achieved only at considerable expense, it seems likely that considerations of cost effectiveness will become more important in justifying future programs.

Quite apart from any scientific advances that would increase the *accuracy* of weather forecasts, there exists an immediate potential for improving their *usefulness*. The basic concept involved in achieving this "improvement in utility" already exists. Further development work is needed, however, to determine how to provide and present the required probability information to the user. Especially lacking are techniques for assessing the prior uncertainty of hazardous or disaster-type weather. Also required are individual studies in operations research, to provide the user with information concerning his operational risks—the costs and losses involved in weather-sensitive activities. Many individuals and organizations in the federal, state, and private sectors are unaware of the magnitude of protectable weather losses that could be alleviated, if not eliminated, by making use of decision-making principles.

Another development aimed at improving the cost effectiveness of the weather services in this country is a new communications system called Automation of Field Operations and Services (AFOS). This system, started in 1977, when completed will consist of a series of minicomputers, a national communications circuit, and television information displays, which will greatly facilitate the immediate relay and distribution of warnings of tornadoes, flash floods, and other important weather events to the key weather service offices for prompt dissemination to the appropriate civil agencies and to the public.

GLOBAL CLIMATE AND CLIMATIC FLUCTUATIONS*

Most human activities are influenced in one way or another by climate. We know from the study of history that the growth and nature of early civilization, the patterns of migration, and the development of agriculture were determined in large part by the available rainfall and the severity of winter temperatures. And our dependence continues, as today's patterns of food production, commerce, and natural resource management are markedly influenced by the climate's vagaries. Climate also plays an important role in many matters of national policy and is therefore a factor in international relations.

With our dependence on climate acknowledged, it is sobering to realize that we actually understand very little of the dynamics of climatic change. The variation of climate is one of the most challenging problems in the atmospheric sciences, and its solution will require the combined efforts of meteorologists, oceanographers, glaciologists, and

*Prepared by W. Lawrence Gates, Oregon State University.

geophysicists and the use of all available methods of inquiry. Already the demand for scientific information on this problem far exceeds the supply of answers; and as man's potential to influence the climate increases, a much better understanding of climate will become a matter of some urgency.

Climatic Change

Man's attitude toward climate seems to have been dominated by either of two views: one, that the climate has always been the same or, two, that nothing can be done to alter the course of a changing climate in any case. We may see that the first view is incorrect by recalling that as recently as 18,000 years ago much of North America and Europe were under vast ice sheets and that the desert regions of the world were once the sites of tropical vegetation. Evidence of such widespread changes comes from a variety of paleoclimatic sources, such as ocean and lake sediments, tree rings, glacier ice, soils, and fossil pollen. The layered structure of such records serves as a natural recorder of climatically dependent biological and chemical processes and may be made to yield indirect or proxy measures of past temperatures and rainfall, for example.

Even more recently, proxy data and scattered historical records have identified a cold period, generally between the years 1400 and 1850, which has been referred to as "the Little Ice Age." Indirect data have shown that the earth has undergone many major and minor climatic changes in the past, and although we do not yet understand their causes, there seems little reason to expect that they will not continue in the future.

The climate is also capable of fluctuating on shorter time scales. The winter of 1972–1973, for example, was cool and wet in the western United States and relatively mild and dry in the eastern part of the country. It was also unusually warm in the Ukraine, which had a direct effect on the harvest of winter wheat produced in that region. Nearly every season brings notable extremes of weather in one part of the world or another, and were these changes to persist from year to year they would constitute a change of climate. Droughts in portions of Africa, Asia, and the Near East are examples of such climatic fluctuations, whose tragic impact on the peoples of these regions is all too well known.

Recorded data show that since about 1940 the average temperature of the air over a major portion of the globe has been slowly falling with a cooling of approximately 0.6°C. We also know that during the period from about 1890 to 1940 the northern hemisphere underwent a gradual warming of over 0.8°C. It has been speculated that this warming was the result of the progressive enrichment of the atmospheric carbon dioxide concentration from the use of fossil fuels and that the more recent cooling trend is the result of the supposedly now dominant effect of atmospheric aerosols, which increase the reflection of solar radiation. On the other hand, it is possible that these temperature trends are natural fluctuations, as yet not significantly influenced by man. We simply do not know.

From historical records we do know that similar climatic fluctuations have previously occurred over at least portions of the earth. The time around A.D. 1000 was one of unusual warmth in the North Atlantic and European regions, and was characterized by relatively mild winters over a period of several hundred years. This coincides with the period of Viking settlement and exploration of Greenland and North America. Looking further back, we know that the earth has experienced at least four major ice ages in the last half million years. The most recent of these began to wane in North America about 16,000 B.C., and the Great Lakes were formed as the ice sheet retreated and melted. How long the current worldwide cooling trend will continue is now the subject of much speculation.

The mere possibility that man might be able to change the climate in some way is of great importance. There is evidence that a relatively small change in solar radiation, for example, might induce widespread climatic changes. And, as already noted, man's alteration of the earth's surface and his injection of larger and larger amounts of waste heat and chemicals into the atmosphere raise the possibility that we may already be changing the climate, possibly in irreversible and undesirable ways.

The Physical Basis of Global Climate

The problem of climate presents us with an extraordinarily complex physical system. The atmosphere, of course, is involved on all scales from the smallest local variations to the large-scale, global circulations of the atmosphere as a whole. It has been long recognized that the oceans are also a crucial part of the global climatic system. The surface waters of the ocean supply a great deal of heat to the atmosphere, both through the evaporation of water vapor (whose latent heat is later released in the atmosphere) and by direct turbulent conduction from the sea surface.

Also involved in the climatic system are the world's ice and snow masses, whether in the form of seasonally varying sea ice and surface snow cover or in the form of the much more slowly changing mountain glaciers and continental ice sheets. The earth's land areas are also involved through their elevation, which influences the circulation and temperature of the air, and through their surface character and vegetal cover, such as desert, forest, cropland, or urban development. Over such surfaces the energy and water balances are profoundly different and may serve to either intensify or ameliorate the surrounding fluctuations of climate.

All four media—the atmosphere, hydrosphere, cryosphere, and surface layer of the earth—are involved in an intricate system of interactions, which is what makes the study of climatic change so complex. We have today only the barest outline of how this system works, and only a few of the important physical processes have received adequate attention.

If we regard climate as the average in some sense of the various elements of weather, then the physical basis of climate is basically the same as that responsible for weather. The primary source of energy is, of course, the sun's radiation, and the transfer, transformation, and dissipation of this energy by the global circulation of the atmosphere and ocean involves all parts of the climatic system. Figure 1 graphically indicates some major processes involved in the climatic system. In these processes, a key role is played by water substance in its various phases and by the transfers of momentum, heat, and moisture through the surface boundary layers. The maintenance of the large-scale pattern of climate is also dependent on the transfers by convective processes, as evidenced, for example, by the close relationship in low latitudes between sea-surface temperature and the rain that falls in convective showers. Over the ocean, the evaporation–precipitation cycle exerts a strong influence on the salinity of the surface water, with the saltier water generally found in regions where evaporation is high and rainfall is low. Over land, however, the hydrologic cycle's interactions with the surface are more complex, since part of the precipitation is re-evaporated, part is absorbed into the system of groundwater, and part is used to nourish the surface vegetation. When the precipitation falls as snow and accumulates, the surface's reflectivity may be greatly raised and the surface energy balance accordingly changed.

Since each of these processes has a characteristic natural variability associated with the weather, the derived climate will also have a certain variability, depending on the time and space averaging used. To distinguish such variability from climatic change itself, it is useful to introduce the concept of a *climatic state*, which we define as the average statistics over a specified period (and in a specified region) of

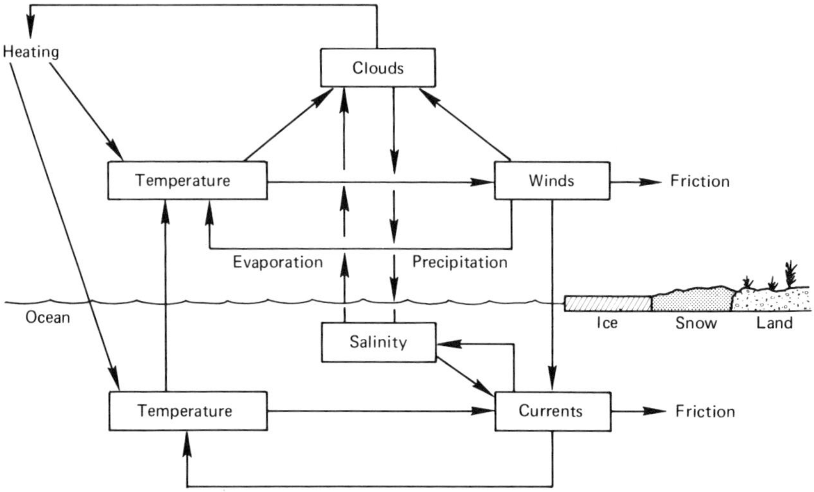

Figure 1 The major physical components and feedback processes responsible for the maintenance of climate.

the complete set of atmospheric, hydrospheric, and cryospheric variables characterizing the system. A *climatic variation* may then be defined as the difference between two climatic states of the same type; and we may speak, for example, of specific monthly, seasonal, or decadal climatic states and of climatic variations over similar time periods. Unfortunately, the climatic data presently available are insufficient to describe many of the important variables and processes involved in even the most recent climatic variations, and the situation becomes progressively worse as older climates are examined.

It is possible that climatic variations may occur as a result of changes in the intensity and position of the various atmospheric processes and feedback mechanisms, even though external or boundary conditions remain fixed. If the latter, such as the sun's radiation or the earth's configuration, are also changing, then these may serve as additional causative factors of climatic variation. (See Figure 2.) Of critical importance over time scales of years to centuries is the oceans' high heat capacity, which may permit them to act as a climatic flywheel confining the fluctuations of climate within narrower limits than would otherwise be the case. Not only do the large-scale ocean currents participate in the global transport of heat toward the poles, but vertical overturning and the newly discovered, medium-scale oceanic eddies may also play significant roles.

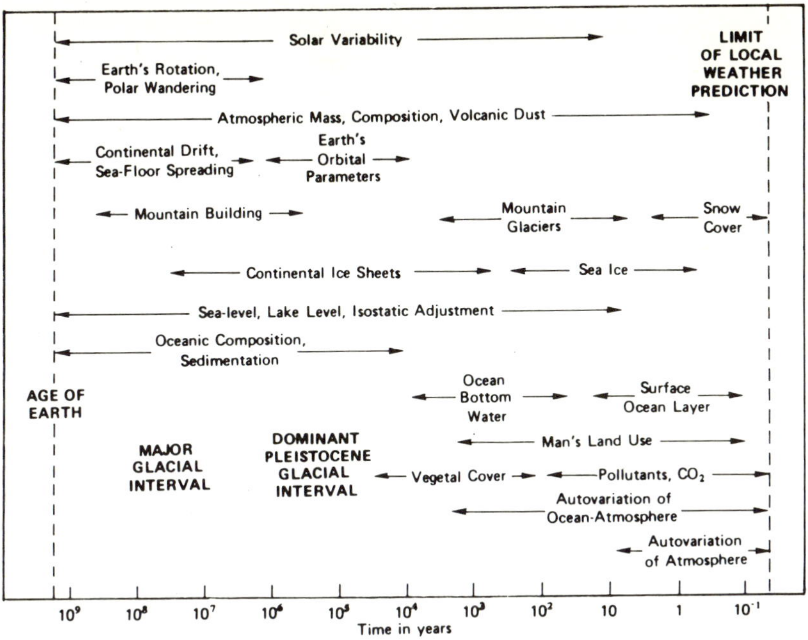

Figure 2 Characteristic events and possible causative factors of climatic change (from Committee on Atmospheric Sciences, *Understanding Climatic Change: A Program for Action,* National Academy of Sciences, Washington, D.C., 1975, p. 22).

Climate Models

Over the past 25 years, meteorologists have developed mathematical models of the atmosphere for the purposes of weather forecasting. These models are based on the fundamental hydrodynamic and thermodynamic equations and, when properly programmed for solution on a high-speed computer, can be made to yield useful forecasts one or two days in advance. The accuracy of the forecast decreases markedly for longer periods. This limitation on the accuracy of weather forecasting is due to the unstable and nonlinear character of the atmosphere and to our inability to portray mathematically all of the interacting components accurately. One of the major goals of the Global Atmospheric Research Program is to obtain adequate worldwide observations of atmosphere and oceans and improved understanding of atmospheric processes. It is hoped that these investigations will make it possible to make accurate forecasts one to two weeks in advance. Even with a complete knowledge of the system, however, the necessarily limited accuracy of the numerical prediction techniques themselves would sooner or later render the predictions inaccurate.

If such a situation exists for short-range forecasts, what hope is there for forecasting the longer-range changes such as are involved in climate? Our hope is that while we cannot tell what the changes will be in a particular place on a particular day, we may be able to say something of the *average* character of the weather, which is, after all, a useful definition of climate. Very simplified models indicate that a small change in the sun's radiation, for example, may be able to cause extensive glaciation of the earth, and we obviously should check such a prediction with the best models available.

By designing global numerical models similar to those used for weather forecasting, and by extending their integration over much longer periods of time, the average worldwide distribution of the various climatic elements can be simulated. By then comparing the simulated distributions of temperature, pressure, wind, and rainfall, for example, with those actually observed in the atmosphere, we can determine how well the model has depicted the climate.

In general, the average large-scale distribution of the sea-level pressure is reasonably well simulated by such models, including the subtropical high-pressure cells and the higher-latitude centers of low pressure. The models also predict the worldwide distributions of temperature and wind at several levels in the atmosphere, together with the cloudiness, precipitation, evaporation, and the heat balance at the earth's surface. The resemblance of the distributions of these simulated climatic elements to those observed is generally satisfactory, although there is evidence of systematic model errors. This is particularly true in the case of the heavy rainfall in the tropics, which most models simulate with too great an intensity. This miscalculation is related to the models' difficulty of accurately portraying local convective showers, which are smaller than the spacing between the locations used in the calculations.

Numerical Experiments and Research

While useful predictability on climate time scales has not yet been demonstrated, a start has been made in a program of numerical experiments in which deliberate changes are made in the initial conditions of the climate model, and the model's solutions are examined to discover the consequences in terms of climate.

In one such experiment, the incoming solar radiation was reduced by a few percent, and the simulated atmosphere displayed the expected cooling. By the end of a 60-day integration, there were no changes that appeared to presage an ice age. As mentioned earlier, however, some theories indicate that such an event would eventually occur were the sun's radiation permanently reduced in this way.

In other general circulation experiments, the temperature of the ocean in midlatitudes was changed by several degrees in order to represent the large-scale pools of anomalously warm or cold water that are sometimes observed. During the first month or so of simulation, this change increased the evaporation and heat flux into the air and lowered the surface pressure over the warmer water. When extended to two months and beyond, such experiments show less clear results, and further experimentation is required to establish the nature of the relations between ocean surface temperature and continental climate that have been suggested by empirical studies.

One of the difficulties in interpreting numerical climatic experiments of this sort is determining which simulated changes are significant and which are not. We noted earlier the lack of predictability present in the weather-forecasting problem and made the hypothesis that there might, nevertheless, be some predictability in the long-term averages. There does appear to be some, but we are not entirely free of the problems that plague the weather forecaster. It seems that any small change—whether it represents a deliberate change of boundary condition (such as sea-surface temperature) or an uncertainty in the data at some point—can to some extent generate anomalies of simulated climate.

For example, if an experiment were to be carried out in which the only alteration was to change the temperature by 0.01°C at one location in the world, the 60-day averaged solution from an atmospheric model would differ from an experiment without the initial change. The solutions would not be greatly different, to be sure, but the changes of wind and temperature would be several meters per second and several degrees, respectively, in scattered regions around the world.

The problem would be to find those regions, if any, in which it could be stated with confidence that a significant change had occurred and which changes were just the reflection of the inevitable "noise" in the model arising from the unpredictable fluctuations of the weather. If the experiment was indeed only the alteration of the *last decimal* in a number stored by the computer or the introduction, say, of a random number field representing the uncertainty of the observed state, there would be no reason to expect a significant change in the model climate. But, if the experiment involved, for example, *large* changes of the albedo over the entire globe, or the removal of all mountains from the North American continent, there would be a reasonable expectation of producing significant climatic changes. Even these, however, would contain an element of uncertainty, and before such experiments can be carried out usefully, we must learn how to separate the climatic signal from the climatic noise. The design of statistical procedures to solve this problem is a goal of current research and is a prerequisite to the successful numerical simulation of climatic change with global circulation models.

In addition to the problem of experiment analysis, further research must be directed to those aspects of the models that are known to be deficient. Of these, the absence of an interacting ocean is perhaps the most important, and extensive efforts are now under way to develop models of the coupled ocean–atmosphere system. In such models, the ocean would respond to the exchanges of momentum and heat across the sea surface, raising the level of complexity (and it is hoped also the realism) of future climatic experiments. Such models appear to offer the greatest opportunity for further understanding the problem of climate, as they alone bring together in a quantitative fashion the diverse elements of the system.

Other important current research is concerned with the models' treatment of the atmospheric surface boundary layer and the surface mixed layer in the ocean. It is through these back-to-back boundary layers that the climatic system receives most of its heat and through which most of the system's kinetic energy is in turn lost by turbulent friction. Like convection, these features are not well resolved in climate models and must be represented in a parametric or indirect way.

In order to gain the maximum possible insight from numerical experiments, it will be necessary to use a variety of modeling approaches. In addition to the global circulation models discussed earlier, there are a number of more simplified models, in which various statistical or empirical approximations are employed in order to reduce the amount of calculation and to deal more directly with the statistics of climate. Although such models may be integrated for extended periods of time, the necessary calibration against more detailed models and against observation has not yet been completed. Once such models for the coupled ocean–atmosphere system are available, they will make it possible to explore a wide variety of possible climatic scenarios.

We must also recognize that the one- or two-month-long simulations that have generally been carried out so far with global circulation models are calculations of specific climatic states and need not be representative of the climate averaged over longer periods. If we are to make meaningful statements regarding the long-term effects of changes in parts of the climatic system, we must simulate at least several years' time in the calculations. This will be expensive in terms of computer time and will require access to the highest speed and capacity machines available.

Outlook for the Future

The need for greater understanding of the global climate is given a sense of urgency by the seeming inevitability of future climatic changes, be they natural or anthropogenic. At present, we know far too little about such changes and are thus ill-prepared either to project or to cope with their consequences. Among the many specific questions to be answered are: If we were to load the atmosphere with dust, for example, or to remove the Arctic sea ice, what long-term effects would this have on the world's climates? What long-term climatic effects would the global operation of a fleet of high-altitude supersonic transports have? And what will happen if man continues to clear forests, to pollute the atmosphere, and to emit waste heat from his expanding energy consumption?

To acquire this knowledge, not only must we construct improved climate models, we must begin the systematic collection and analysis of the global climatic data necessary to calibrate the models and to support the necessary diagnostic and empirical studies. This, in turn, will require the dedication of significant financial resources to climatic research on a long-term basis and new national and international programs to provide the necessary coordination.

There must be greater efforts to learn about climates of the past through the use of geochemical, dendrochronological, and other innovative approaches. Imaginations should be set free to speculate and hypothesize on causes and effects.

Further research should enlarge our understanding of the dynamics of climatic change and assist in the development of an adequate theory of climate. It might then be possible to specify beforehand what the long-term climatic consequences of specific actions will be, or at least to put limits on the consequences, and it may even be possible to predict the general character of the year-to-year climatic fluctuations themselves. Without such a theory, we cannot discern confidently the results of man's interference with the environment from the background of natural changes. The stakes are certainly very high, and the prospect is a worthy scientific challenge.

4

Environmental Preservation and Management

AIR QUALITY*

Over the last few decades we have learned a great deal about air pollution. The well-known disasters in Donora, Pennsylvania, in 1948 and in London in 1952, and others as well, in which large numbers of people died following several days of heavy atmospheric contamination, were dramatic illustrations of the consequences of atmospheric contamination. The much more common episodes of polluted air in larger urban centers such as Los Angeles, New York, Chicago, Tokyo, London, and Moscow have made it clear that if we do not maintain, and in some cases restore, air quality, society will continue to pay dearly for our neglect. For a long time, the price has been high in terms of the health of humans and animals, losses in agricultural production, property damage, and the intangible losses when we no longer can see the beautiful sights of distant mountains and deep blue skies.

More recently, it has become evident that man-made pollutants can have crucial indirect effects on the atmospheric environment. One of the vital questions of the day is whether the emissions of carbon dioxide and particles into the air are influencing the global climate.

*The major portion of this section was contributed by Donald H. Pack, Air Resources Laboratory, National Oceanic and Atmospheric Administration.

Other important uncertainties need to be resolved. For example, to what extent can human activities change the compositions of the upper atmosphere? Will the nitrogen oxides of high-flying airplanes and the chlorofluorocarbon gases from spray cans and refrigerators significantly reduce the concentration of ozone in the layers some 50,000 to 150,000 feet (15 to 45 km) high? To what extent would such changes lead to increases in skin cancer and other serious biological problems?

There are many more questions we can ask about the quality of the atmosphere, how it is changed and the consequences of the changes. Much has been learned about these matters. We have made many measurements and have made significant advances in the chemistry and physics of the atmosphere. At the same time, as it usually is the case, the more we know about nature, the more we discover we still have to learn.

Properties of Air

It is usual to call "air" a gas; in fact it is a mixture of gases. Mostly it is nitrogen (78 percent by volume) and oxygen (21 percent by volume), plus a great many other constituents in small or even trace quantities. In addition to the many components whose concentrations are fixed, there are a number of important gases whose concentrations are small but highly variable. Notable among these are water vapor, carbon dioxide, ozone, and many others that are products of human activities and can be put into the category of pollutants. Some of the most prominent are listed in Table 2.

TABLE 2 Average Surface Concentrations of Certain Gaseous Pollutants

Gas	Approximate Average Concentrations (Parts per Million)
Carbon dioxide (CO_2)	331
Carbon monoxide (CO)	0.1–0.5
Hydrocarbons	Less than 0.01
Ozone (O_3)	0.02–0.1
Nitrogen dioxide (NO_2)	0.001
Sulfur dioxide (SO_2)	0.0002
Fluorocarbons (CCl_3F)	0.0002
(CCl_2F_2)	0.00013
Methyl chloride (CH_3Cl)	0.002–0.0005

Some of these gases are of great interest and deserve particular attention. Before examining them, however, we should also note that there are huge numbers of tiny solid and liquid particles in the atmosphere.

Atmospheric Particles

Excluding clouds, which are composed of water droplets or ice crystals, most of the particles in the atmosphere are soil or smoke or substances formed by chemical reactions of various gases in the atmosphere. Certain metals from smelting or combustion activities can be toxic in sufficiently high concentration. As an example, one can cite lead particles from automobile exhausts.

Atmospheric particles are generated in many ways: grinding, impaction, combustion, break up of liquids, condensation, coagulation. Most natural particles are produced by the wind (dust storms and sea spray), biotic exudation, and volcanic activity. Many of man's activities generate particles, particularly those involving motor vehicles and smoke stacks.

The composition of particles has not been measured in any systematic and continuing monitoring program outside of urban areas. However, inside of cities, they have been shown to encompass almost all of the metals and, often through absorption onto the surfaces of other materials, hundreds of chemical compounds. One of the more widespread materials measured has been lead (Pb), much or most of which has come from the use of leaded gasoline. The combustion process encourages the formation of many small particles with radii less than a few micrometers [1 micrometer (μm) = 10^{-6}, or one millionth of a meter (m)]. Inferences about the origin of particles (i.e., whether they arise from man's activities) can sometimes be made by noting the relative amount of a substance in the earth's crust and in the air. For example, measurements made at the South Pole indicate that the metals zinc, copper, cadmium, lead, tin, and selenium are present in quantities that can only be attributed to man's activities.

The greatest single problem in the measurement of particles is their great range of sizes and the small mass concentrations. The sizes range over about 5 orders of magnitude from near 10^{-9} m to 10^{-4} m.* No one measurement technique can be used over a size range of much more than 1 or 2 orders of magnitude. Fortunately, however, the larger particles, those greater than about 10 μm in radius, fall out of the

*Particles ranging in size from one billionth to one ten thousandths of a meter.

atmosphere rather rapidly by gravitational settling and are, therefore, primarily of local and regional interest. On the other hand, particles in the smallest size range, less than 0.01-μm radius, do not settle out and are not removed readily by rain and snow processes. These particles, unless they aggregate to form larger ones, stay in the atmosphere for periods amounting to a few tens of days.

The latest estimate of particle production available was for 1968 and was prepared for the United Nations Conference on the Human Environment in Stockholm in 1972. The total global production was estimated at about 2.6×10^{12} (2.6 trillion) kilograms (or kg) per year. Of this amount, about 40 percent is believed to be less than 5 μm in size; hence, relatively long lived. Man's contribution to the total was estimated to be about 16 percent, or near 4×10^{11} (0.4 trillion) kg per year. The ratio of particles produced directly (dusts, metals, sea salts, etc.) to particles created by the conversion of gas to particulate (e.g., sulfur dioxide and ammonia, combining to form an ammonium sulfate particle) is about two to one.

Data on the concentration of particles are numerous, but a few typical values will suffice. To provide a perspective, the U.S. primary standard for total suspended particulates is 260 micrograms (μg) per cubic meter (m^3) for a 24-hour period.

From Table 3, it will be seen that larger cities have larger values on an annual basis. But almost any location can have a high 24-hour value.

TABLE 3 Examples of Total Suspended Particulate Values during Recent Years (Micrograms per Cubic Meter)

Location	Highest 24-Hour Value	Annual (Geometric) Mean
Florence, Alabama	1830 (highest reported in publication)	Not Available
Anchorage, Alaska	1393	90
Los Angeles, California	412	114
St. Louis County, Missouri	354	42
New York, New York	157	73
Chicago, Illinois	130–463	60–164
Hilo, Hawaii	51	25
Yellowstone National Park, Wyoming	24	9

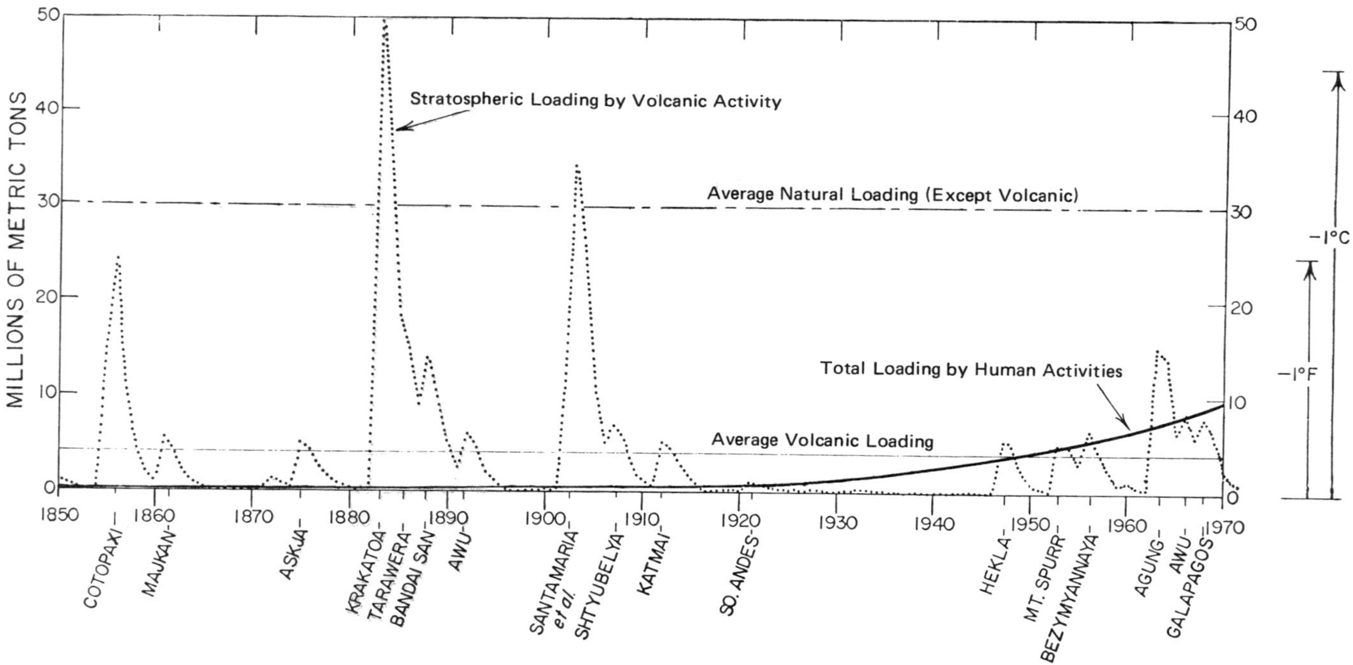

Average annual global atmospheric particle loading, 1850–1972 by volcanic activity (stratospheric loading only, dotted curve), by human activity (heavy solid curve), and by all natural sources, other than volcanic (assumed constant at 30 million tons, dash-dotted line). The estimated calibration of volcanic loading curve in terms of planetary temperature influence is shown at right (after J. M. Mitchell, Jr., 1975: *The Changing Global Environment*, p. 164. S. F. Singer, Edit., D. Reidel Publishing Co., Dordrecht, Holland).

Several stations in Oklahoma, not included in Table 3, had 24-hour values well above 1000 μg per m^3, quite possibly caused by dust storms. The lowest known value of suspended particulates was obtained at the NOAA Geophysical Monitoring Observatory at 3400 m elevation on Mauna Loa in Hawaii. Here the daytime values are from 5 to 10 μg per m^3 falling to about 0.1 to 0.5 μg per m^3 at night.

What can be said about the trend of particles in the atmosphere? Analyses by scientists of the Environmental Protection Agency show that in the 13 years between 1960 and 1973, cities became cleaner and especially that the higher values of suspended particulates occurred less often. In the rural areas, the evidence is not so clear; the particle collections showed little change, but solar radiation attenuation data and satellite photographs indicated some regional deterioration. Measurements, over the oceans, of concentrations of minute particles and of atmospheric electrical properties indicate a "plume" of pollution off the eastern coasts of Asia and North America but little or no long-term changes in the central ocean areas. Since the attenuation of solar radiation by solid and liquid particles could have an important bearing on climate, it is necessary to monitor their concentrations locally, regionally, and globally and to follow the trends carefully.

In spite of the previous comments, which were directed toward long-term steady trends, there is unequivocal evidence that atmospheric particles have attenuated solar radiation for considerable periods of time. The sources of these particles are large volcanic eruptions injecting large amounts of particles and gases into the stratosphere. While smaller eruptions may create vast local clouds of dust and gas, unless they penetrate into the stable stratosphere above the cloud-bearing layer, they will fall out or be removed too quickly to have any effect much beyond the immediate area. In contrast, massive explosions inject huge amounts of material to great heights. The explosion of the Krakatoa volcano on the Sunda Strait (about 6°05′ S, 105°30′ E) in August 1883 ejected about 1.8×10^{16} m^3 of material into the atmosphere! Not all, or even most, of the material went into the stratosphere, but some did and to heights above 25 km. D. Deirmendjian of the Rand Corporation has analyzed two other catastrophic volcanic explosions, Mt. Katmai, Alaska, on June 6, 1912, and Mt. Agung on Bali on March 17, 1963. He estimated the volume of material blown into the stratosphere as follows: Krakatoa, 3×10^9 m^3; Katmai, 1.34×10^9 m^3; Agung, 9×10^8 m^3 (3 billion m^3, 1.34 billion m^3, and 0.9 billion m^3, respectively). There were few quantitative measurements on the first two events and, surprisingly, no really organized study of even the Agung eruption. However, considerable data have been

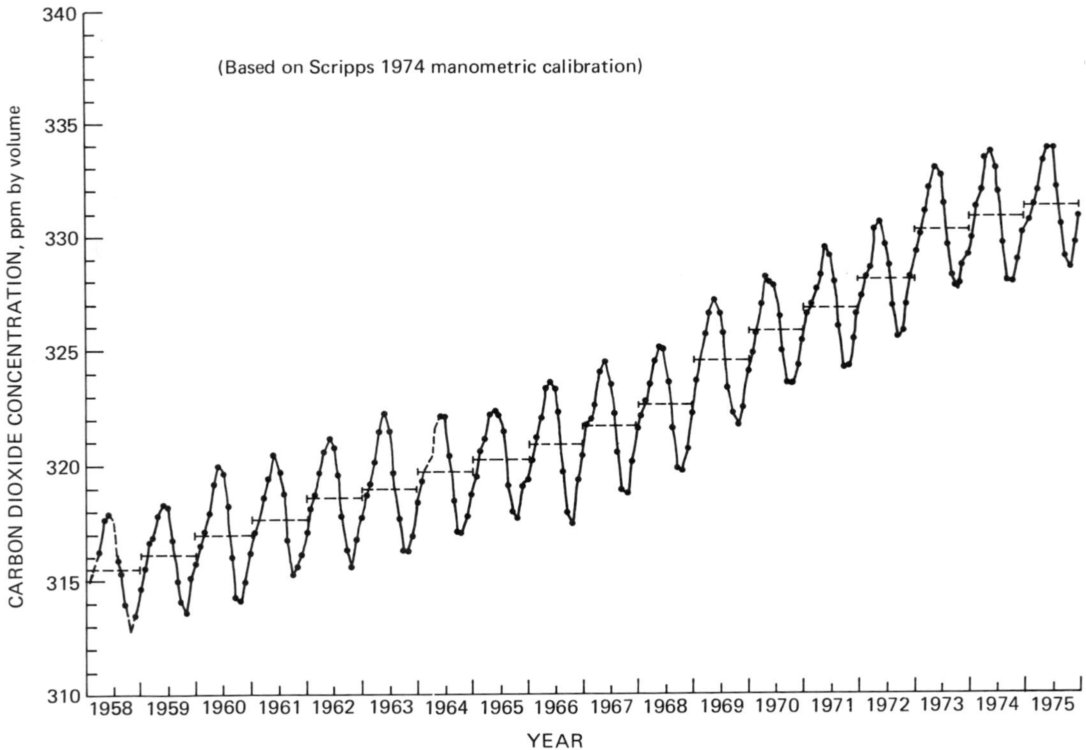

Figure 3 Seasonal and annual variation in the carbon dioxide (CO_2) content of the atmosphere, 1958–1975. (Courtesy, NOAA)

accumulated after the Agung event. All the explosions increased the turbidity of the atmosphere for a few years. The total solar radiation (direct and diffuse) did not appear to be much changed, but the direct solar beam was attenuated by nearly 2 percent.

In October 1974 an eruption of the volcano Fuego, in Guatemala, again injected significant amounts of material into the stratosphere. The wind transported the cloud over Hawaii shortly after the Mauna Loa Observatory lidar (a lasar radar), installed specifically to monitor stratospheric dust layers, had become operational. As of March 1975, the layer was still easily observed by the lidar, although the layer was thicker and less sharp than in October immediately after the eruption.

A cooperative study of the stratospheric dust layers is under way to map the changes with time in both middle and subtropical latitude zones. Simultaneously, there is an intensive study of related data on solar radiation and temperature structure. It is essential to learn what effects atmospheric particles are having on weather and climate.

Atmospheric Gases

Carbon Dioxide and Carbon Monoxide Carbon dioxide has been in the earth's atmosphere for at least two billion years. Prior to 1850, its concentration in the atmosphere was about 280 parts per million by volume (ppm). It is now about 331 ppm. This increase can be attributed to the CO_2 released when coal, gas, and oil are burned and is therefore one of the few direct and unequivocal pieces of evidence that man is altering the atmosphere. This alone, however, would not be enough to require the intensive studies that have been made, since CO_2 is relatively inert and is not involved in any important chemical reactions in the atmosphere. However, its properties are such as to permit the penetration of shortwave solar radiation through the atmosphere without hinderance, while intercepting, through absorption, the longer-wave infrared radiation from the earth's surface. This process, if it acted alone, would lead to a warming of the lower layers of the atmosphere. As noted in the discussion of climate, it is not that easy to infer the consequences of CO_2 absorption on the earth's climate.

Figure 3 shows the longest continuous record of CO_2 now available. The measurements, collected on the top of Mauna Loa were initiated during the International Geophysical Year under the direction of C. David Keeling of the Scripps Institution of Oceanography.

Although the upward trend of CO_2 concentration is quite evident, there are still mysteries. The changes up until about 1968 could be tied closely to the available data on global annual burning of fossil fuel.

However, after that date the annual changes at Mauna Loa (and at a South Pole station) have been more erratic and proportionately greater than the increases in the use of fossil fuels. Are the data unrepresentative? This does not seem likely, but only two points on the globe are not many to represent the entire earth. A small decrease in the uptake by biota or a small rise in ocean temperature (all things being equal, the colder the ocean the more CO_2 it can absorb) could produce the observed effect. Keeling has recently proposed that the fractional amount of CO_2 released into the atmosphere, which will remain there, will increase significantly above the present 50 percent as we move into the next century. This would lead, of course, to a more rapid increase of the CO_2 concentration in the atmosphere than has occurred in the past.

A major expansion in the number of global observations of CO_2 is under way with the United States increasing the number of stations to four, a Canadian program, measurements by Australia, New Zealand, and Sweden, and probably more to come. All data are tied together with common calibrations and data-exchange agreements.

Carbon monoxide does not present a threat to the global climate as does CO_2. On the other hand, it is hazardous to health, and it is fascinating in the number of surprises that it has given atmospheric chemists over the last few years. In 1972, combustion processes in the United States alone are estimated to have emitted 4.6×10^{10} (46 billion) kg of CO. Global anthropogenic production for the same year was about ten times this much. However, it has been found that some soil bacteria can rapidly reduce the concentrations of CO. On the other hand, measurements over the oceans indicate that CO is produced by microbiological processes and that the rate of creation is highest in nutrient-rich waters. This enables the oceans to be a significant source of this gas.

Observational data are mostly confined to urban areas, since the measurements of the very low values in "clean" air are quite difficult to obtain. Open-air urban values range from 2–10 ppm, while values inside automobiles in traffic jams have been observed to reach 400 ppm. In contrast, values over the Atlantic Ocean are about 0.18 ppm and near 0.03 to 0.05 ppm over the ocean in the southern hemisphere. While there is now no observational evidence to show a growth on CO concentration, neither are there data to show it constant. Although this gas, in high concentrations in cities, can have deleterious effects on humans, there is no reason for believing that it is having any significant widespread effects on the biosphere or atmosphere.

Sulfur Dioxide Any discussion of sulfur dioxide gas by itself is necessarily incomplete, since it should be evaluated in terms of the total sulfur cycle in which it plays only a partial and generally short-lived role. It is included here because it may be produced in very large amounts in certain smelting operations and by the burning of fossil fuel with high sulfur content. For this reason, sulfur dioxide has long been singled out as an important industrial pollutant, and much effort has gone into the measurement of ground-level concentrations and into engineering developments to reduce these levels.

Each year about 10 million kg of SO_2 are emitted into the atmosphere by man-made sources. Natural production is less than a tenth of that and comes primarily from volcanic eruptions. These emissions are highly variable from year to year and are believed to range as low as 1 percent up to 20 percent of the man-made production. Levels of SO_2 in the atmosphere have been found to be as low as 0.0002 ppm in remote locations and as high as 0.5 ppm near uncontrolled heavy industrial emitters. The U.S. standard for a 24-hour maximum is 0.14 ppm.

Sulfur dioxide has a short lifetime in the atmosphere. It oxidizes to sulfate at rates that are now known to vary as a function of humidity and the pressure. Certain metals seem to catalyze the reaction. In clean, dry, and relatively aerosol-free air, SO_2 may have a lifetime of 2 to 3 days; while in a humid, dirty air mass containing submicrometer-sized metals (e.g., manganese) half of the sulfur dioxide may be converted to sulfate in about 30 minutes. Recent research is also pointing up that sulfates, presumably in the form of particles small enough to be breathed into the respiratory system, pose a more direct health hazard than sulfur dioxide gas.

It is known, both in Europe and in the eastern United States, that rainwater has become increasingly acidic over the past years. While the rainfall pH* at Mauna Loa Observatory is not far below the neutral value of 7, east of the Mississippi River rain consistently has a pH value of about 5. A year's data from a small network in the Washington, D.C., areas shows that in 1974 the rain pH averaged 4.1. Both sulfur dioxide and sulfates have been indicted as a major factor (about 60 percent) in accounting for acid rain. The evidence indicates that important ecological effects of sulfur dioxide emissions extend more than a thousand kilometers away from the sources.

*pH is used to indicate both acidity and alkalinity, having a numerical scale of 0–14 with 7 representing neutrality, numbers less than 7 increasing acidity, numbers greater than 7 increasing alkalinity.

Ozone Ozone (O_3) has been of interest since the discovery early in this century that ultraviolet solar radiation at wavelengths shorter than about 3000 angstroms (one angstrom is a unit of wavelength of light equal to one ten-billionth of a meter) did not reach the surface because it was being absorbed by ozone. The cause and effect relationship between radiation and ozone was first presented in 1930 by the late Sidney Chapman, who showed that ozone was created by the photochemical action of ultraviolet radiation on oxygen with the consequent depletion of the radiant energy in this band. The bulk of the global ozone is, in contrast to most trace elements, produced high in the atmosphere. On the average, the maximum concentration is near 25 km, although it varies with season and latitude.

Ozone is a reactive gas and participates readily in chemical and photochemical reactions. Since it is of basic importance in providing a shield against the solar ultraviolet radiation and since the energy absorbing properties also dominate the stratospheric temperature, its importance to human welfare is not measured by its comparatively small amount.

Ozone also is found near the earth's surface and can be an irritant gas that, in concentrations much exceeding 0.08 ppm, affects the human and animal breathing apparatus. Natural background ozone levels in remote areas range from 0.02 to 0.04 ppm depending on the downward transport of ozone through the atmosphere. Concentrations may reach higher values in storm systems as the transport from aloft is temporarily accelerated. Over rural, vegetated countryside, recent measurements show levels of 0.08 to 0.1 ppm. Within cities, especially those with large automobile populations, ozone is created by the reactions of automobile exhaust products under the action of sunlight, specifically the longer-wave ultraviolet light that reaches the earth's surface. City values of oxidant, much of which is ozone, have reached average maximum values of about 0.2 ppm and individual peak values of 0.8 to 1.0 ppm.

As a part of the interest in the ozone layer, there has been a resumption of the study of the natural variability of its amount. By far, the largest body of data on total ozone amounts has been obtained by special spectrophotometers measuring the relative radiation intensities in carefully selected wavelength pairs. There are about 50 such observation stations over the globe. In addition, the Soviet Union operates about 30 observation stations using a somewhat different instrument. The latest analysis of these data shows significant natural variations of the order of several percent. These findings, while important in themselves, also indicate that detection of any ozone change that is a small fraction of the total, say less than 1 percent per year, will be difficult or impossible until quite a few years have passed.

We have the paradox that high values of ozone near the earth's surface demonstrably create undesirable effects and great effort is being exerted to reduce the concentrations by controlling the release of materials that react to create it. On the other hand, the existence and maintenance of high values in the stratosphere are essential to the maintenance of life. Until relatively recently, the existence of the high layer was taken for granted, since there seemed no way in which interference could occur. It was not until about 1970, just as the debate concerning the environmental effects of supersonic transports flying in the stratosphere was coming to its climax that it was realized that a gas appearing copiously in the exhaust of the engines of these planes—nitric oxide—also reacted catalytically with ozone and was present in the stratosphere naturally in amounts probably sufficiently large to significantly reduce ozone concentrations.

Recently two comprehensive studies have been made of the effects of high-flying aircraft and their nitrogen oxide emissions on the ozone layer. They concluded that the effects would be very serious if large numbers of aircraft propelled by today's jet engines were to fly in the stratosphere. According to the report of the Climatic Impact Committee of the National Research Council,* 500 American SST's flying at 19.5 km in the airlanes of the northern hemisphere, each one 5 hours per day, would deplete the ozone abundance in the northern hemisphere 16.5 percent and in the southern hemisphere 8 percent. This result is somewhat larger, but not significantly so, than the conclusions of the Department of Transportation's Climatic Impact Assessment Program.

The atmospheric models used to calculate the effects of nitrogen oxides on ozone were one dimensional, that is, they considered vertical transport of gases to be governed by a diffusion equation. This procedure is certainly a reasonably well justified one above 100 km. In the lower atmosphere, it is necessary to take into account the effects of turbulent air motions. This has been done by means of a simple, one-dimensional eddy-diffusion model. There are several problems with this approach, nevertheless, it gives results that appear to be reasonable. Also, it is useful as a technique for first-order predictions of atmospheric phenomena that deserve to be measured. There is clearly a great need for tractable three-dimensional transport models

*Climatic Impact Committee (1975). *Environmental Impact of Stratospheric Flight: Biological and Climatic Effects of Aircraft Emissions in the Stratosphere.* National Academy of Sciences, Washington, D.C.

that can include gas chemistry along with gas motions. No such models exist at the time of this writing.

Chlorofluorocarbons and Ozone Recognition of the effects of nitrogen oxides on the ozone layer set scientists thinking about other possible anthropogenic sources that might lead to ozone destruction. It was found that chlorine atoms in the upper atmosphere could lead to a reduction in ozone concentrations and that large quantities of chlorine-bearing molecules are being emitted into the lower atmosphere. The principal ones are chloromethanes and chlorofluorocarbons, sometimes called fluorocarbons. These substances came into widespread use some two decades ago because the sulfur dioxide and ammonia used as working gases in mechanical refrigerators were unpleasant and could be hazardous. New and better compounds were needed and were developed within the chlorofluorocarbon family. The materials are essentially inert and do not combine with anything, they are not inflammable, are essentially nontoxic to man and animals, and are almost completely insoluble in water. It was found that fluorocarbons were well suited for use in aerosol-type dispensers.

The compound most widely used in refrigeration systems is CCl_2F_2. A second fluorocarbon, CCl_3F, was developed for use in the aerosol spray cans. The characteristics of each compound seemed to be well suited to its application. Over the recent past, there has been a large increase in the use of chlorofluorocarbons. The production of CCl_2F_2 was 5.9×10^7 (59 million) kg in 1958. In the same year, 2.3×10^7 (23 million) kg of CCl_3F was manufactured. By 1973, the latest year for which data are available, the world-wide production had grown to 4.7×10^8 (0.47 billion) kg of CCl_2F_2 and 3.1×10^8 (0.31 billion) kg of CCl_3F, with about half of these amounts being produced in the United States.

Chlorofluorocarbons have been found in the lower atmosphere in concentrations of 50 to 100 parts per trillion (ppt) and appear to be increasing, although the evidence is shaky. The observed total amount is about the same as the quantities of fluorocarbons that have been produced since 1961. About 2350 metric tons of CCl_2F_2 and 3500 metric tons of CCl_3F had been produced by the end of 1973, and this would work out to 80 ppt if it were thoroughly mixed in the troposphere and if its lifetime were infinite, that is, if there were no sinks to diminish atmospheric concentrations. Lately evidence turned up that there may be an appreciable sink in the ice of the South Pole. Furthermore, another very stable chlorocarbon, CCl_4 (carbon tetrachloride), has been found to be mysteriously present in the atmosphere in the mixing ratio of about 100 ppt. Other interesting Cl compounds, such as CH_3Cl (methyl chloride), are being discovered at frequent intervals.

In any event, in the absence of sinks in the earth's crust, the ocean, or the troposphere, the fate of the chlorocarbons and fluorocarbons (CCl_3F, CCl_2F_2, and CCl_4), because of their great chemical stability, is to diffuse up to the barrier of the tropopause, penetrate it along with convective clouds or through gaps in the tropopause, as the mixing ratio builds up in the troposphere, then diffuse upward in the stratosphere until finally near 30 km they are far enough through the ozone layer so that they can be exposed to the ultraviolet solar radiation normally absorbed by the ozone at lower elevations. This radiation is capable of freeing chlorine atoms and effect the destruction of O_3. The great height at which chlorine atoms are released makes them particularly dangerous compared with the nitrogen oxide molecules emitted at 20 km by jet engines. The reason is that the duration of molecules in the stratosphere is longer when they originate at 30 km than when they are injected at 20 km. The lifetime of a chlorine atom in the stratosphere is 5 to 10 years. It would be expected that the destruction of ozone would occur long after the release of chlorofluorocarbons into the lower atmosphere.

Early calculations indicated that if the release of chlorofluorocarbons continued to increase at the present rate of 10 percent per year until 1995, the maximum ozone depletion would occur in the year 2001, when it would be about 14 percent.

On the basis of recent findings, the percentage increase in erythemal ultraviolet irradiation at the ground and the incidence of skin cancer in humans is about double the ozone percentage decrease; these projections are alarming indeed.

Since the early estimates were made, there have been a series of redeterminations of chemical reaction rate constants and a number of measurements in the stratosphere of constituents that are important in ozone chemistry. New processes involving interaction of the chlorine and nitrogen catalytic systems have been postulated. The result is that the predicted ozone depletion has varied from the early value to amounts as small as 1 or 2 percent. The best present estimate is that the effect is probably about half as large as originally predicted, but it is too early to assume that all the factors involved in this complex chemical–dynamical system are understood.

Already an intensive measurement program is under way, designed to detect the concentration of chlorine compounds in all levels of the stratosphere. Remote-sensing techniques in the infrared and microwave regions of the spectrum and direct *in situ* spectroscopic detection of the relevant gaseous species from balloon or rocket-launched instruments are the favored present methods of attack.

These examples of the discovery almost by chance and almost belatedly of threats to the ozone caused by man's activity has alerted us to the need of understanding the atmospheric–biospheric system involved. It is essential that we be able to anticipate future dangers and be in a position to react wisely and expeditiously when they are recognized.

WEATHER MODIFICATION*

Effective techniques to control the weather, if employed wisely, could serve society well. Quite obviously there would be substantial savings if fogs over airports could be dispersed on a reliable basis and allowing commercial airliners to land instead of having to fly to a distant, alternate airport. Statistics are not needed to show that the ability to increase rainfall during a drought would be beneficial. Who could argue that the weakening of violent storms and the consequent reduction of fatalities and storm damage are not desirable goals? If the net benefits of a weather-modification technology exceed the net costs, there is little doubt that it would be used extensively. It is essential to realize, however, that as is the case with many complex technologies, it is difficult to evaluate the costs and benefits particularly when the body of relevant knowledge is inadequate. In recent years, social scientists have been analyzing the value of weather modification. In this discussion we will focus on the scientific aspects of the problem.

Over the years, the science of weather modification has been punctuated by promising advances and liberally sprinkled with major disappointments. Social, legal, and environmental considerations, which in the 1940's and 1950's were given only passing attention, in more recent years, have begun to play more prominent roles in any consideration of this subject. Such questions as, "Who owns the rain?" and, "Who has the right to decide whether a tropical storm should be seeded?" and "Should weather modification techniques be included in a country's military arsenal?" are much more than philosophical questions.

Before examining some of the broad questions dealing with the social consequences of weather modification, it is in order to examine the scientific and technical aspects and assess what has been accomplished in the past and what might be achieved in the future.

*Prepared by Louis J. Battan, University of Arizona.

Scientific Basis of Weather Modification

Although history is rich with attempts to change the weather by incantations, rain dances, the ringing of bells, and firing of cannons, the most significant steps in the development of scientifically reasonable methods to modify clouds began in 1946. A series of important investigations was carried out at the General Electric Research Laboratories by Irving Langmuir and his associates, particularly Vincent Schaefer and Bernard Vonnegut.

They discovered that if finely crushed dry ice or a smoke produced by the burning of the compound silver iodide were introduced into certain clouds, ice crystals could be produced. This occurred only when the clouds were *supercooled*, that is, they were composed of liquid droplets even though their temperatures were below the usual freezing point of 0°C. The failure of the water to freeze can be attributed to the fact that the cloud droplets that formed by the condensation of water are made up of relatively pure water. It has long been known that many clouds in the atmosphere are supercooled, often to temperatures below −10°C.

Photomicrograph of crystals of silver iodide (AgI). (Photographed by Roger J. Cheng, State University of New York—Albany)

After a series of laboratory experiments, Langmuir and his colleagues showed that by seeding supercooled clouds with *ice nuclei* (i.e., substances such as dry ice or silver iodide particles) they converted them to ice crystals. The crystals, in turn, grew large enough to fall to the ground either as individual crystals or aggregates of crystals in the form of snowflakes. In some atmospheric conditions, the heat released as the supercooled water freezes increases cloud buoyance, which may cause additional cloud growth.

Over the last three decades, a great deal of work has been done, in many parts of the world, to develop improved cloud-seeding techniques, which can change effectively certain characteristics of the weather. Most of the time and resources have gone into the following activities: (1) dissipating fogs over airports, (2) increasing rain or snow, (3) decreasing the fall of damaging hail, (4) reducing the frequency of cloud-to-ground lightning, and (5) reducing the intensity of hurricanes.

Fogs over Airports

When a persistent fog covers an airport, the costs and inconvenience to the airlines and the passengers can be quite high. In 1968, it was estimated that the loss of revenue to the users of a major airport by one fog occurrence was as high as $100,000. With the advent of large, wide-bodied airplanes, the costs of a major fog episode over an airport might be as high as $500,000.

Fortunately, ice-nuclei seeding is effective in causing supercooled fogs to be dissipated as ice crystals form, grow, and fall to the ground. A variety of procedures have been used effectively for many years at airports in the United States, France, and the Soviet Union.

Unfortunately, about 95 percent of the fogs occurring over airports in the United States are not supercooled. Instead, they are composed of water droplets having temperatures above 0°C and are known as "warm fogs." Ice-nuclei seeding is useless in such conditions.

During World War II, Allied pilots operating out of English airports frequently were grounded by heavy fog or were forced to make treacherous landings at hidden airports. Dramatic success in fog clearing was achieved with the development of a system called FIDO (Fog Intensive Dispersal Of). It consisted of lines of pipes with small holes through which aviation fuel flowed and was burned. The heat warmed the air and caused the evaporation of fog droplets close to the ground. After the war, it was found to be too expensive for use at commercial airports.

Over the years, many schemes have been conceived for the dissipa-

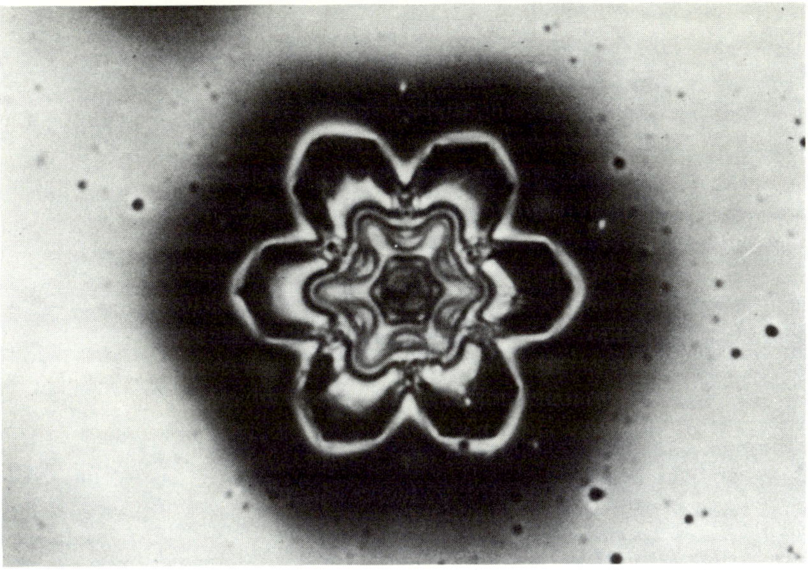

Ice crystal nucleated by lead particle from engine exhaust. (Photographed by Roger J. Cheng, State University of New York—Albany)

tion of warm fog. They have included the dispersion of various salt solutions into the fog, the hovering of helicopters, the heating of air over runways, and the use of laser beams to evaporate the droplets. The last scheme, proposed by Soviet scientists, would work with small volumes of fog, but it seems doubtful that it would be practical for clearing airport runways. At the present time, a technique, which in its most sophisticated version in France is called *Turboclair*, appears to have the most promise, particularly at airports handling large airplanes. Turboclair is a modern version of FIDO. It involves the use of a series of eight jet engines installed underground along the edges of the runway. The hot gases from the engines warm and stir the fog and cause the droplets to evaporate. This technique is being investigated in the United States and the Soviet Union.

More Rain or Snow

Over the last few years, newspaper headlines have reported the consequences of inadequate rainfall during the growing season. Droughts in the grain belts of a number of countries in 1972 dealt a serious blow to the food supplies of the world. The results were rising prices for everyone and starvation for many. In mid-1975, wholesale prices for grains were substantially lower than they were in the middle of 1974. To a certain extent this is a reflection of optimism about the weather. If it is favorable, food production in the United States will reach record levels, and, for a time, concern about adequate supplies is likely to subside. Let us hope that if that does occur, concern is not replaced with complacency. The needs for food in the world will continue to increase as population continues to climb.

Shortly after it was demonstrated in 1946 that supercooled clouds could be converted to light snow, which fell to the ground, the search began to find effective means for increasing the quantity of precipitation. It was known for a long time that raindrops often originate as ice crystals and snowflakes that melt as they fall through warm air near the ground.

There is no doubt that ice-nuclei seeding of some supercooled clouds can cause more snow or rain than would have occurred if the cloud had not been seeded. There still is considerable uncertainty about how much more precipitation can be produced in any particular meteorological circumstance and whether the increases are socially significant.

A detailed discussion of the difficulties in arriving at definitive answers about the efficacy of rain augmentation experiments will not be attempted here. The chief difficulties arise because of the great natural variations in precipitation from place to place and time to time because of our inability to predict expected rainfall amounts accurately. A great deal still must be learned about natural precipitation mechanisms. In these circumstances, it is necessary that, in designing an experiment, the investigators make use of the best available information about the physics of the problem and employ sound statistical procedures in the planning, conduct, and evaluation of the experiment. Over the last 30 years, there have been a large number of cloud-seeding operations in many countries that have not followed that prescription. As a result, in most instances, it is difficult or impossible to establish whether seeding had any effects. In too many cases, the operators assumed that seeding would increase rain or snow and acted accordingly. Lacking effective controls from which to judge what would have occurred without seeding, evaluations are almost hopeless.

Fortunately there have been, in several countries, a number of carefully designed and executed cloud-seeding programs, which serve as the basis for generalizations about the efficacy of cloud seeding. In addition, a few commercial cloud-seeding operations have been conducted over the same area, following the same procedures, for enough years that they allow some inferences on the likely effects of seeding to be made.

In 1973, a report of the National Academy of Sciences* concluded that ". . . ice-nuclei seeding can sometimes lead to more precipitation, can sometimes lead to less precipitation, and at other times the nuclei have no effect, depending on the meteorological conditions." It was noted that changes of perhaps 10 to 30 percent might be expected in some circumstances. The most convincing evidence to support this conclusion came from the seeding of snow-producing, winter clouds over the mountains in the western United States. Other important data bearing on the quoted conclusions, came from experiments on the seeding of clouds in Australia, Switzerland, Israel, Florida, Arizona, and the midwestern United States.

Unfortunately, little is known about the specific meteorological conditions that govern how cloud seeding will work. Over the Soviet Ukraine and Florida, there have been a number of experiments yielding fairly convincing proof that the rain from certain types of summer clouds can be substantially increased. Nevertheless, in both places, it has not been shown yet that during the summer growing season, it is possible to increase significantly the quantity of rainfall over an area some tens of kilometers across.

It should be noted, however, that the number of first-rate cloud-seeding investigations have been quite few. They require excellent scientific and statistical talent, sufficient resources to do the job right, and sufficient time to gather enough data to permit meaningful conclusions. Fortunately, a few such projects are currently under way or in the planning stage under the auspices of the National Oceanic and Atmospheric Administration and the Bureau of Reclamation.

Over the last decade, there has been growing concern about the effects of seeding on the precipitation falling over regions downwind of the seeded target area (which might be only some 30 to 50 km on a side). Jerzey Neyman and his associates at the Statistical Laboratory of the University of California at Berkeley have been particularly active in the study of this problem. They have concluded that ice-nuclei seeding may cause substantial increases or decreases of rain at distances as much as 200 km or more beyond the boundaries of the target area. In every cloud-seeding operation the possibility of effects downwind of the primary area of interest should be recognized and provisions made for estimating their magnitudes.

In times of droughts, in regions where rain or snow are the principal sources of water, farmers, the managers of hydroelectric utilities, and municipal officers often look to cloud seeding for assistance. Unfortu-

*Committee on Atmospheric Sciences (1973). *Weather and Climate Modification: Problems and Progress*. National Academy of Sciences, Washington, D.C., p. 4.

nately, in such circumstances clouds are scarce and opportunities for seeding are few in number. Nevertheless, at some times, in some places, it might help, while at others the results could be detrimental. If cloud seeding is employed, provisions should be made to evaluate the results.

There is a pressing need to reduce the uncertainties about the effects of cloud seeding on rainfall. It is essential to make a national commitment to the search of a technology for reliable precipitation augmentation.

Reduction of Hail Damage

In many parts of the world, hailstorms do extensive damage to vegetation. The grain fields of the United States suffer perhaps $500 million in losses in an average year. Over northern Italy and the southern parts of the Soviet Union, fruit orchards and vineyards are devastated on a regular basis. In East Africa, the tea plantations suffer periodic destruction by hailfall. Many other examples could be cited.

Although much remains to be learned about the nature of hailstorms and the factors governing their growth and decay, certain facts have been known for many years. Hailstones are particles of ice ranging in diameters from 0.5 cm to greater than 10 cm. Extremely large ones, approaching the size of grapefruit, have occurred on rare occasions. Most often, damaging hailstones are from 1 to 2 cm in diameter and fall in large numbers.

In order for hailstones to form, it is necessary that there be fairly strong updrafts and large quantities of supercooled water. The rising air inhibits the fall of the stones toward the ground. In the cold regions of the cloud, supercooled water droplets collide with the stones and the accreted water freezes. The hailstones continue to grow as long as they remain in the supercooled part of the cloud. When they fall into warm air, the stones begin to melt; those only a few millimeters in diameter may melt completely and reach the ground in the form of raindrops.

Favorable conditions for hail formation are found in thunderstorms, particularly those with strong sustained updrafts that tilt and twist as the air ascends through the storm. Such conditions occur more frequently over the Great Plains of the United States than they do over tropical regions such as Florida.

The idea that the growth, size, and structure of hail could be modified by means of ice-nuclei seeding has been around more than 20 years. Over the last decade or so, the evidence that it might be possible has been increasingly enticing.

Most attempts to modify the quantity of damaging hail are based on certain assumptions. First, it is assumed that within a thunderstorm, the quantity of water available for hail growth is essentially constant. Second, it is assumed that by introducing ice nuclei, it is possible to produce many more hailstones than would have occurred if the storm had not been seeded. In the event that the number of hailstones is greatly increased, each one might be so small that it would melt while falling through warm air near the ground.

Over the last couple of decades, there have been many hail modification *operations* in many countries, most of them involving the seeding of clouds by means of silver iodide or lead iodide nuclei. The results have been inconclusive for the most part. In most operations, the absence of controls makes it impossible to judge what would have happened had there been no seeding. Even in the cases of hail-modification experiments that incorporated statistical controls, the results have been mixed. In some instances there was more hail following seeding, and in other instances there was less hail.

In the late 1950's Soviet scientists in the Caucasus region began an intensive attack on the hail problem. By the early 1960's, scientists at the High Altitude Geophysical Observatory began to use 100-mm artillery guns for firing ice nuclei into potential hailstorms. Another group in the Republic of Georgia in the southern Caucasus began using rockets for the same purpose. From the beginning, they reported spectacular success in reducing hail occurrence and hail damage. In 1963, seeding was being carried out over about 300,000 acres of farmland, mostly fruit orchards and vineyards. By 1974, other seeding groups were in action, and the area under seeding had increased to about 10 million acres, according to Yu. Sedunov, Director of the Institute of Experimental Meteorology near Moscow. He claimed that hail damage to agriculture was being reduced by amounts between 70 and 95 percent. Soviet scientists estimated that benefits from hail seeding amounted to about ten times the costs.

Such glowing statistics are impressive but surprising in light of experiments and hail research outside the Soviet Union. Before accepting them as proof that hailstorms can be modified, it would be appropriate for other scientists outside the Soviet Union to test the Soviet hypotheses of hail suppression. Any procedure that can increase agricultural output at costs that are far below the benefits cannot be overlooked.

The National Hail Research Experiment (NHRE) was established in the United States in 1969 with the hope that it could test the validity of the Soviet claims of suppressing hail damage. The artillery and rocket-seeding techniques used by the Soviets could not be duplicated over the NHRE test area in eastern Colorado. Instead, seeding was done from airplanes. After three years of systematic seeding experiments, NHRE scientists were not in a position to offer convincing judgments on the efficacy of the Soviet methods and have been seeking greater understanding of the nature of hailstorms and how to modify them.

A joint Swiss–Italian–French experiment is being planned to test Soviet seeding rockets on hailstorms in Switzerland. The rockets and launchers have been purchased from the Soviet Union, and attempts are being made to duplicate procedures used in that country. The Swiss experiment is expected to be a reliable test of Soviet results.

In conclusion, many atmospheric scientists find themselves in a position of being optimistic about the possibility that hail damage can be suppressed but still lacking the body of data that would constitute the basis for a hail-suppression technology. A research program appropriately well staffed and funded should clarify the situation, but it is likely to require another five to ten years of work.

Lightning Suppression

During the 1960's, researchers of the U.S. Forest Service conducted experiments to test if the frequency and intensity of cloud-to-ground lightning could be reduced by seeding thunderstorms over Montana with ice nuclei. One set of data suggested that the number of fire-causing lightning flashes could be reduced, but the tests were not pursued to the point where such a conclusion could be regarded as proven. Other similar tests in other parts of the world have led to similarly inconclusive results.

In more recent years, scientists of the National Oceanic and Atmospheric Administration have investigated the degree to which large numbers of small, metallic fibers can change the electrical properties of thunderstorms and possibly reduce lightning frequency. As with other lightning-suppression research, this work has not been carried far enough. Early encouraging results have not been adequately tested.

It is disappointing that support for research on lightning modification has been meager and uncertain. At this time, inadequate funds have reduced research in this area almost to the vanishing point.

Lightning does enormous damage to many forms of property; it kills more people in an average year than do tornadoes; it causes great inconvenience by interrupting power distribution systems and communication links. Modification experiments should be viewed as means to learn about the properties of thunderstorms and the lightning they

Massive lightning strike from summer thunderstorm near Boulder, Colorado. (Courtesy, NCAR)

produce. Such information would be of great value in designing lightning-sensitive systems and structures. Furthermore, a lightning-suppression technology can provide protection when large numbers of people congregate in the open for public meetings or sporting events or when lightning sensitive operations such as the launching of manned spacecraft are about to take place.

Hurricanes and Typhoons

When the winds in a tropical cyclone over the Atlantic Ocean exceed 119 km/hour (74 miles/hour) it is given the name *hurricane*. Similar storms in the Western Pacific are called *typhoons*, while those in the Indian Ocean are called *cyclones*. Regardless of the name, the storms are the same. They develop over warm ocean water at tropical latitudes. In an intense hurricane approaching the United States, the winds blow counterclockwise at speeds that can exceed 300 km/hour (186 miles/hour) over a nearly circular area perhaps 100 km in diameter. At the center of the storm is the "eye," a relatively calm region, typically some tens of kilometers in diameter and sometimes mostly free of clouds. The eye is bordered by towering thunderstorms—bands of storms that extend outward more than 200 km in the form of giant cloud spirals. As an airplane flies outward from the eye, it encounters rapidly increasing winds, which reach a peak at a distance perhaps 50 km from the center of the storm. At greater distances, the wind diminishes gradually reaching perhaps 30 km/hour at the edge of the storm.

In the late 1940's Irving Langmuir and his associates at General Electric Research Laboratories reasoned that by seeding hurricanes with ice nuclei it should be possible to modify them. In October 1947, he conducted such a seeding experiment and observed that the storm subsequently followed a peculiar track. There was considerable debate among atmospheric scientists over whether the storm's behavior was caused by the cloud seeding, but a subsequent analysis indicates that the hurricane apparently started changing its path before it was seeded.

In the early 1960's the U.S. Government established Project Stormfury for the purpose of studying hurricanes and testing whether they could be weakened by artificial means such as cloud seeding. Because of the need to conduct experiments over the open ocean where life and property were not in jeopardy, there were few experimental opportunities. The most encouraging ones occurred in the late summer of 1969.

On August 18, 1969, Hurricane Debbie had peak winds of 182 km/hour (113 miles/hour) at the altitude of 3600 m (about 12,000 ft) when it was heavily seeded with silver iodide dropped into the storm from specially equipped airplanes. Five hours after the seeding ended the winds had dropped to 126 km/hour (78 miles/hour), a 31 percent fall. On August 19, no seeding was done and the storm reintensified. On August 20, with peak winds of 183 km/hour (113 miles/hour), seeding resumed. Some five hours after it ended, peak winds had diminished to 156 km/hour (97 miles/hour), a drop of 15 percent. These results were in conformance with the expectations of the hypothesis being tested. Analyses of wind data, as well as many other measurements made in and around the hurricane, led to increased optimism that hurricanes might be weakened by means of ice-nuclei seeding. Complicated, but nevertheless still somewhat scientifically incomplete, mathematical models of hurricanes led to the conclusion that seeding of towering supercooled clouds outside the ring of peak winds should lead to a decrease in the speed of those winds.

Following the encouraging results of the Hurricane Debbie experiments, there was substantial optimism that hurricane seeding might diminish peak winds and hence reduce hurricane damage without an accompanying reduction of total rainfall. At the same time, it was recognized that the experimental results were extremely limited in number and the changes in wind speed in Hurricane Debbie might have occurred by chance, rather than as a result of seeding. It was hoped to repeat the tests on other full-fledged hurricanes. Unfortunately, none occurred over the experimental areas in 1970, 1971, or 1972. The investigation of an anomalous storm called Hurricane Ginger in September 1971 shed little light on the effects of seeding.

In 1973, the Project Stormfury field experiments were discontinued. One of the chief reasons was that the older airplanes employed in the tests had to be replaced. New airplanes are being obtained and instrumented. Present plans call for Project Stormfury to be reinstated in 1977.

In view of the tremendous influence of hurricanes on the lives of so many people, it is important that we know more about them, how to predict them, and how to modify them beneficially. It should be recognized that, although hurricane damage normally gets the headlines, the benefits of hurricane rainfall are in many instances very high. In some cases, the value of hurricane rainfall may even exceed the losses, but different individuals are usually involved. Most of the damage is expected to occur close to the coast, while the beneficiaries are likely to be agricultural interest and municipalities further inland. At any rate, in evaluating the efficacy of a hurricane seeding program, such factors must be taken into account. According to the hypothesis being investigated by Project Stormfury scientists, heavy ice-nuclei seeding outside the region of peak winds should reduce the peak winds and wind and storm-surge damage without, at the same time, reducing the total rainfall from the hurricane or changing the storm's path.

The potentialities of hurricane seeding are quite substantial, and the resumption of the tests must be regarded as an important field of investigation.

Social Consequences

It is surprising, considering the extent of cloud-seeding operations in the United States, that there has not been a greater public response for or against such activities. It is true that over the last couple of decades or so there have been perhaps a couple of dozen legal actions taken, but that seems a small number when viewed against the overwhelming number related to other environmental matters. In some specific cases in Pennsylvania, Texas, California, Colorado, and most recently in Rapid City, South Dakota, there has been considerable public interest and even tensions on matters related to who had the right to "change the weather." Certain legal authorities have proposed that weather-modification activities be subject to regulation by the federal government. At this time, it is necessary that all weather modifiers report their activities, but there are no federal licensing or control provisions. At the moment, about half of the states have laws dealing with the licensing or regulation of weather-modification activities. They vary widely. Some states require that prospective weather modifiers give evidence of meteorological or engineering competence. Others merely require reports.

At periodic intervals, bills dealing with weather modification are introduced in Congress, and occasionally hearings are held, but there has been only one substantial legislative action since 1968 when Congress discontinued the National Science Foundation's responsibility for weather modification. In 1971, the National Oceanic and Atmospheric Administration was assigned responsibility for collecting information on nonfederal government weather-modification activities. About one year later, federal weather-modification programs were included in these reporting requirements.

Over recent years, there has been growing interest in establishing international controls on the use of weather-modification techniques for military purposes. A 1971 report of the National Academy of Sciences* recommended that the United States sponsor a United Nations resolution dedicating weather modification to peaceful purposes. Senator Claiborne Pell of Rhode Island has led a drive to

*Committee on Atmospheric Sciences (1971). *The Atmospheric Sciences and Man's Needs: Priorities for the Future*. National Academy of Sciences, Washington, D.C., p. 79.

establish international agreements prohibiting the use of "geophysical weapons" in warfare, and Congress adopted a resolution to this effect. In 1974, President Richard Nixon and General Secretary Leonid Brezhnev of the Soviet Union signed an agreement to study means to overcome the dangers of the use of environmental modification techniques for military purposes. Later the same year, the Soviet Union introduced a resolution in the United Nations General Assembly that would prohibit the use of environmental and climate-modification techniques for other than peaceful purposes.

Clearly, there are many legal, social, ecological, and political questions bearing on the subject of weather modification. Unfortunately, they are complicated by the limited ability of atmospheric scientists to specify, at this time, what meteorological changes can be made by means of cloud seeding or other modification procedures.

It is unfortunate that at a time when beneficial modification of the weather has become clearly important, support for research has diminished. The national investment decreased from $20 million in 1973 to about $15 million in 1975. Such a trend must be reversed if any substantial progress is to be made. Weather modification represents one important option in overcoming shortages of water for farmlands, power companies, and municipalities. It offers an important approach in dealing with devastating storms that not only impose high costs on society but also cause large numbers of injuries and deaths.

Although there have been advances over the last few decades, progress in weather-modification research has not been fast enough. The problems are difficult ones, and the potential benefits are great. The time is right to conduct a vigorous, sustained national research program.

THE UPPER ATMOSPHERE*

The outer limit of the earth's atmosphere is difficult to define. Near the ground, gaseous material exists in abundance; i.e., a cubic meter of air has a mass of about a kilogram. This quantity, called the air density, decreases gradually with height until, at altitudes of several hundred kilometers, it is measured in terms of molecules and atoms per cubic meter.

Up to about 80 km, the relative amounts of the various constituents of air change very little. Mostly, air is composed of molecular nitrogen (78 percent by volume) and oxygen (21 percent by volume). At greater

*Edited version of manuscript by Thomas M. Donahue, University of Michigan.

altitudes, the absorption of ultraviolet radiation from the sun causes the dissociation of the atomic and molecular constituents of air. The zone of ionized gas extends through a depth of several hundred kilometers. At still greater heights, one encounters the magnetosphere, which is contained within a vast region circumscribed by the boundary between the incoming solar wind and the magnetic fields of the earth.

The term "upper atmosphere" is used to specify that part of the earth's atmosphere above about 10 km, the average position of a relatively horizontal surface called the tropopause. It is a level where temperatures are a minimum and separates the stratosphere from the lowest layer of the atmosphere. The study of the upper atmosphere and the processes taking place there is called *aeronomy*.

As will be noted later, research on the many fascinating phenomena of the upper atmosphere presents great challenges to the imagination and ingenuity of chemical and physical scientists with broad ranging interests. It will also be seen that the state of the upper atmosphere and its changes can have important effects on such practical problems as worldwide communication and power distribution. It also has been proposed that air motions and weather in the lower atmosphere can be influenced by events in the high atmosphere.

Before examining the practical aspects of upper-atmosphere research, it is appropriate to review briefly the nature of and the scientific problems confronting aeronomers. We start at the outer limits of the atmosphere and move toward lower altitudes.

Physical Properties of the Upper Atmosphere

Far out in space the solar wind (an ionized gas called a plasma), which is flowing away from the sun, encounters the earth's magnetic field. Upstream in the direction of the sun, ten or more earth radii away, the wind forms a shock wave and compresses the earth's magnetic field on the sunward side. Downwind the earth's magnetic field stretches away in a long tail extending hundreds of earth radii to the moon and beyond. The magnetic field enclosed by the cavity-shaped region formed by the solar wind is populated with energetic, charged particles, mostly protons and electrons, and is called the magnetosphere. There are still major problems in understanding how the solar-wind plasma penetrates this cavity, is heated to tens of millions of degrees, and is injected into low altitudes at high latitudes to cause magnetic storms, auroras, and other interesting and important phenomena. Of a more controversial nature are the intriguing correlations that have been shown to exist between solar-wind variations and the weather. Whether this phenomenon is a statistical fluke over a physically causal relationship between magnetic activity and weather (or vice versa) is a topic of great interest currently.

In the magnetosphere, the extensive high-altitude region of very low gas density and infrequent collisions between ions and molecules, the motion of ions is controlled by their interaction with the geomagnetic field. At high geomagnetic latitudes, that is, in the polar cap above the auroral oval, the ionized gas moves toward the sun. It travels away from the sun at slightly lower latitudes below the polar cap in such a way as to complete its circuit. At still lower latitudes, the plasma rotates with the earth.

The collective energy of the ionized particles trapped in the geomagnetic field occasionally increases dramatically, a phenomenon called a *magnetic storm*. The storm itself is accompanied by a number of substorms characterized by large-scale disturbances in the tail of the magnetosphere downwind from the sun. These disturbances are associated with dramatic increases in the occurrence and intensity of

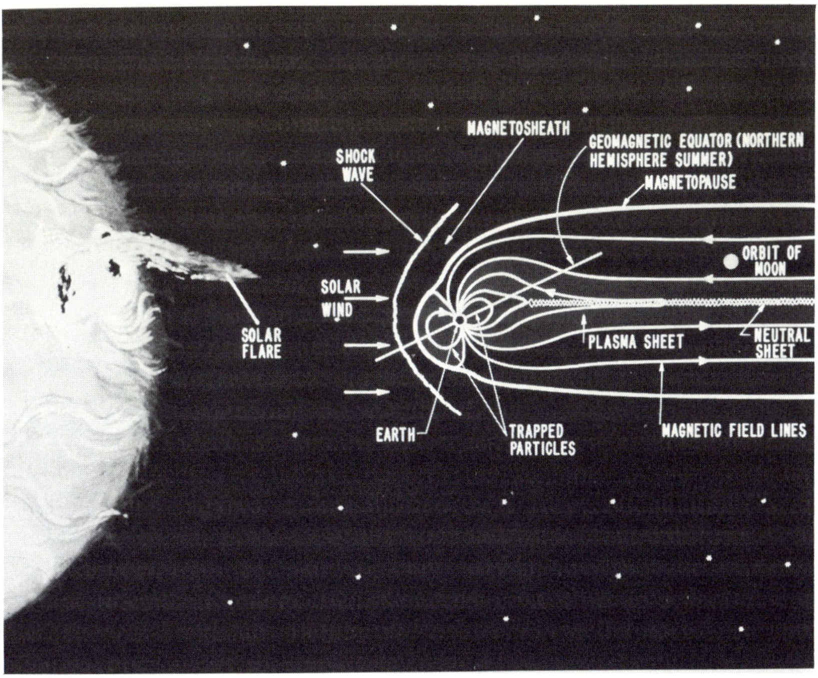

Artist's view of effect of major solar wind upon the magnetosphere of planet Earth. (Courtesy, NOAA)

auroras. The nature of these substorms and the exact relationship between them and the intensification of auroral activity constitute one of the principal unsolved problems of magnetospheric physics.

As a matter of fact, there still are problems in explaining the aurora, identifying the regions in the magnetic tail and the solar wind from which the energetic particles that cause auroras come, and accounting for the mechanisms by which the particles are accelerated. Other key questions are to what extent the circulation of the magnetospheric plasma is coupled with the neutral atmosphere below and what is responsible for the occasional, strong electric fields parallel to the magnetic field lines near the auroral zone.

Major advances in understanding of the magnetosphere are expected over the last half of the 1970's as a consequence of the International Magnetospheric Study. This worldwide research program, under the sponsorship of the International Council of Scientific Unions, calls for coordinated observations from spacecraft, ground-based facilities, airplanes, balloons, and rockets.

The outermost region of the earth's electrically neutral atmosphere is called the exosphere. The base of the exosphere is located about 500 km above the surface of the earth. It is an imaginary surface that separates the part of the atmosphere where collisions between gas molecules and atoms are frequent and the region where this is no longer the case. Above this surface, gas particles, if their kinetic energies are large enough, can escape from earth. Only hydrogen atoms escape in this manner in large quantities. More will be said about this subject later.

Below the exosphere is the thermosphere. Here the sun's electromagnetic radiation, in the x-ray and extreme ultraviolet portions of the spectrum, is capable of ionizing the principal constituent gases of the upper atmosphere, molecular nitrogen (N_2), oxygen (O_2), and atomic oxygen (O). So much energy is deposited in the thermosphere (between about 85 and 300 km) in this way that the temperature of the rarefied gas in this region rises from about 180°C at 85 km to the order of 1000 K at 250 km, even above 2000 K during times of intense solar activity.

In most of the thermosphere, that portion above about 100 km, the dynamical processes that keep the constituent gases of the atmosphere well mixed exist but are weak. Simple molecular diffusion controls the vertical motion, and the lighter molecules tend to be higher than the heavier ones. First N_2, then O, then atomic hydrogen (H) dominates the atmosphere. However, the thermosphere is not a dynamically calm region of the atmosphere. Over the global scale there are large fluctuations in gas motions, temperature, and composition. During geomagnetically quiet times, when there are no magnetic storms, the circulation is driven by the absorbed solar ultraviolet energy. But during storms, auroral processes can set up strong gas flows from high to low latitudes.

As noted earlier, the processes by which energy in the solar wind is transferred to the thermosphere are still poorly understood. Over the next decade, research on this and other important related problems will be carried out by means of a variety of techniques ranging from specialized radar probing of the upper atmosphere from the ground to the Atmosphere Explorer and Electrodynamics Explorer series of satellites.

In the thermosphere, above 90 km, exists the major portion of a vast region of electrically charged atoms and molecules known as the ionosphere. The growth in detailed understanding of the basic ion chemistry of this part of the ionosphere was one of the triumphs of aeronomy during the 1960's and early 1970's. It demonstrated the power of collaboration between laboratory studies of microscopic atomic and molecular processes and field measurements by instruments on space vehicles or ground-based sounders and radar. So far, the lowest reaches of the ionosphere, known as the D region and lying between 60 and 90 km, has resisted determined efforts to achieve a satisfactory understanding of its intricacies. It is a region characterized by strange ionic species, far more positively charged nitrogen oxide than can be accounted for, layers of metallic ions, and a plethora of strange negative ions.

The attention of aeronomers has been drawn, during the past half dozen years, to the stratosphere and mesosphere. As noted earlier, the lowest region of the atmosphere, the troposphere, is separated from the stratosphere by the tropopause, across which vertical mass transport is very slow.

In the stratosphere and mesosphere, dynamical processes compete with photochemistry for control of minor species: ozone, methane and compounds or radicals of carbon, hydrogen, nitrogen, and oxygen. Ozone has already been discussed earlier in this chapter in the section on air quality. Great difficulties are confronting the efforts of atmospheric scientists to produce quantitative models of this middle portion of the atmosphere. The reason is the complexity of the chemistry and the poor understanding of the three-dimensional air circulation. Attempts to model the ozone distribution with one-dimensional transport characterized by a simple vertical eddy diffusion, while producing useful estimates of the dominant chemical processes, are of dubious quantitative predictive value. Actual transport, particularly through the tropopause, cannot be regarded as simple diffusion. Recent measurements of the vertical distribution in the stratosphere of chemical tracers, showing them to vary strongly with time and latitude, attest to the perils of attempting to represent vertical transport by a simple eddy-diffusion model in the middle atmosphere.

Such models have proven useful, however, at higher altitudes, where photochemical reaction rates are rapid. Compared with the troposphere, the stratosphere is very dry, and it is not clear whether the net flow of water is upward or downward across the tropopause. Methane is oxidized to produce water in the upper atmosphere; and above 100 km various hydrogen compounds release atomic hydrogen, which flows upward through the thermosphere and escapes at the rate of about 25×10^7 cm^{-2} s^{-1} (7×10^7 tons per year of equivalent water). Interestingly, it can be shown that the escape flux is determined mainly by the amount of hydrogen in the form of H_2O, CH_4, and H_2 in the stratosphere. The fact that hydrogen escapes at about ten times the rate previously supposed has intriguing implications for theories of the origin of respiratory forms of life on earth. The reason is that the time required for oxygen to build up in the atmosphere as a result of dissociation of H_2O and escape of H is drastically shortened, and it becomes easy to see how an ozone shield would have grown thick enough to permit primitive photosynthetic organisms to exist near the surface of the ocean as long as one billion years ago.

Effects of Upper-Atmosphere Phenomena

During geomagnetic storms, many types of communications systems can experience serious difficulties. The storms can induce large currents in long-distance cables and cause the disruption of switching, amplification, and relay circuits.

When magnetic storms occur, the electrical characteristics of the ionosphere are changed and propagation of radio waves via the ionosphere can be interrupted. Long-distance radio transmissions are carried out by means of ground-to-ground transmissions, depending on reflections from the charged species in the ionosphere, or by a ground-to-satellite-to-ground link. In both instances, changes in the ion concentration, particularly in the lower part of the ionosphere, pose serious threats to effective radio reception.

The large electrical currents that can be induced in electrical lines not only affect telephone and telegraph cables, they also can cause undesirable electrical transients in power distribution systems. The induction effects can trip alarm relays that signal operational problems in power transformers. Such a result can have serious consequences in metropolitan regions. In 1958, a great geomagnetic storm initiated a series of events that caused a blackout in Toronto, Canada. Power companies currently use information of geomagnetic activity in planning their daily operations.

For many years, there has been research relating solar and geomagnetic activity with circulation patterns and weather in the lower atmosphere. The evidence for such relations usually has taken the form of correlations of one or more meteorological factors with some characteristic of the sun or, more recently, some characteristic of the earth's magnetic field. No one has yet demonstrated a physically plausible mechanism by which a perturbation in the upper atmosphere is transmitted to the lower atmosphere. Nevertheless, this intriguing topic continues to get a great deal of attention. The motivation is, at least in part, the hope that such results may shed light on factors governing seasonal and annual weather and climatic fluctuations.

5

Planning and Managing Atmospheric Research

Improved understanding of the earth's physical environment is essential if we are to deal successfully with the problems of food, water, energy, and natural hazard. The atmospheric sciences, in concert with related disciplines, have much to contribute to the nation's quest for this understanding. A number of examples can be given:

- While food production is obviously dependent on agricultural genetics and technology, it is also limited by the seasonal weather, by poorly understood changes in temperature and rainfall that occur over large regions. Production of both food and hydroelectric power conceivably might be increased through techniques of cloud seeding; and these possibilities are likely to grow in attractiveness as water supplies become more critical.
- The energy problem is linked in a variety of ways to the capacity of the atmospheric boundary layer to dissipate heat and to disperse, transform, and transport pollutants. In developing new energy sources we must be prepared to anticipate and to guard against inadvertent effects on human health.
- The effects produced by high-flying aircraft and by chlorofluorocarbons or other stable compounds on the stratospheric ozone shield appear to be cumulative; we must understand these effects long before they reach a level that could be harmful to life.

Prepared by Robert G. Fleagle, University of Washington.

• The loss of life from hurricanes, tornadoes, and floods remains high, because, despite impressive strides in improving warning services, predictions are insufficient in range and accuracy.

In order to reach the improved understanding required by these examples, major research programs are needed. Some of them are under way, and others are being planned. The scientific rationale behind them has been provided in reports of the National Academy of Sciences and of government agencies, and therefore we believe it safe to assert that it has been shown that atmospheric research has much to contribute to the nation's effort to meet the problems of food, energy, and natural hazard. Each of these problems is expected to become more critical in the future than it has been in the past, and consequently the demands on atmospheric scientists for research results will become more urgent. The response of the atmospheric sciences community to these challenges will depend heavily on adequate funding, on the supply of scientific and technical manpower, on effectively managing and coordinating research programs, and on complex decisions at national and international policy levels. The history of the atmospheric sciences shows that these essential components can be assembled, but a close look at some of the major current programs exposes some serious deficiences and indicates that changes in institutional structure and in administrative procedures may be needed to meet the challenges of the future. Such changes are most likely to occur when there is a degree of consensus within the scientific community and appropriate decisions at high policy levels of government.

This discussion addresses some of the institutional and policy issues that are likely to be major determinants of future research programs. As background, this chapter reviews briefly procedures and organizational structures within which policy decisions are made within the federal government. In addition, the federal atmospheric research budget and the research and educational capacities of our universities are examined. Special attention is given to five existing major research programs: the Global Atmospheric Research Program (GARP), weather modification, air quality, inadvertent modification of the stratosphere, and climate change.

THE DECISION-MAKING ARENA

In trying to understand the action in the decision-making arena it is helpful to distinguish three functions or roles requiring participation by scientists that are essential components of the total process. They are

(1) advocacy, (2) objective evaluation and policy advice, and (3) management and interagency coordination. Although from time to time one institution, or even one individual, may play more than one of these roles, there are dangers in this practice, and it is probably a good general policy that the three responsibilities be separate.

Advocacy often has been the role of the scientific community speaking through the channels of the National Academy of Sciences and its National Research Council (NRC). In the atmospheric sciences, the reports of the NRC Committee on Atmospheric Sciences and the NRC Committee for the Global Atmospheric Research Program have been effective in influencing research policy and initiating major research programs. Federal agency scientists and administrators also have been effective as advocates of agency programs, and other groups occasionally have assumed advocacy roles. The five programs to be reviewed later will provide several examples of advocacy, some more successful than others.

The role of objective evaluation and policy advice requires at high levels in the executive and congressional branches of government broad scientific competence, sound judgment, and the capacity to tap specialized expertise wherever it exists. During the 1960's, the Office of Science and Technology and the President's Science Advisory Committee, both headed by the President's Science Adviser, provided objective evaluations and policy advice. Recommendations on program and budget priorities were made to the agencies concerned and to the Bureau of the Budget and, on occasion, to the President. In the late 1960's, this component of the decision-making structure was downgraded and fragmented, and in 1972 it was abolished. Some of these functions were passed to the new Science and Technology Policy Office under the direction of the Director of the National Science Foundation (NSF). However, the close association of this office to NSF made it difficult for the Office to mediate effectively among agencies or to advise the Office of Management and Budget on agency budgets.

During the period 1972 to 1975, when potential crises were recognized to be developing in the fields of climate, weather modification, and modification of stratospheric ozone, special committees were established under the President's Domestic Council. It was widely recognized that this was not an efficient procedure for dealing with new crises as they arose, and efforts were mounted in the scientific community and in the Congress, as well as in the White House, to restore the capability for objective evaluation and policy advice to the President. In May 1976, passage of the National Science and Technology Policy, Organization and Policy Act of 1976 provided for establishment of the

Office of Science and Technology Policy within the Executive Office of the President. The new law also creates a Federal Coordinating Council for Science, Engineering, and Technology and a President's Committee on Science and Technology.

It should be noted also that the National Academy of Sciences, in response to government requests, provides evaluations and policy advice on specific critical problems; to be effective, these must be addressed to administrative levels capable of making policy decisions.

Congress has been involved in the past only marginally in the evaluation of science programs. Although it is responsible for authorization and appropriation of agency budgets, it has not had the technical expertise to match that available to the executive branch. However, Congress is taking an increasing interest in science budgets and in science policy. The Congressional Reference Service of the Library of Congress, the General Accounting Office, and the recently created Office of Technology Assessment provide Congress with competent advice in science and technology. Furthermore, a few Congressional committee staffs have been strengthened by the addition of scientists and engineers familiar with the technical aspects of the agencies whose budgets they review. Congressional interest in coordination and management is illustrated by the introduction in October 1976 of S. 3889, a bill to establish a Department of the Environment and Oceans.

Once a major program has been approved, there must be provisions for the necessary organization, management, and coordination, functions requiring participation of scientists at the policy level. This is more critical in the atmospheric sciences than in many other scientific fields because of the nature of the science. Resource and manpower requirements for research in such areas as weather prediction, air quality, weather modification, and climate can only be met by the joint efforts of all capable groups in government units and in the universities and research institutions. Research on such programs typically requires complex observing systems deployed over large regions, often for extended periods. Intricate methods of analysis are required, often involving collaboration of several technical or scientific disciplines. Such large-scale, complex research programs depend critically on effective organization and management, as well as on adequate manpower and facilities. Deficiencies in management may seriously limit what can be achieved or may even prevent any significant achievement at all.

Recognizing these factors, the Federal Council for Science and Technology (now the Federal Coordinating Council for Science, Engineering, and Technology) in 1959 created the Interdepartmental

Committee for Atmospheric Sciences having three principal functions: (1) to survey and evaluate the national research effort in the atmospheric sciences, (2) to examine the role and activities of federal agencies therein, and (3) to make recommendations for appropriate allocation of responsibilities among the federal agencies. This Committee has served as a forum for interagency discussion and has been an effective device for communication among the agencies involved in atmospheric research. It has compiled informative annual summaries of agency budgets that support atmospheric research. It is important to note, however, that because the Interdepartmental Committee for Atmospheric Sciences has no authority over member agencies, they may launch programs without the Committee's endorsement. As a result, problems of organization, management, and coordination have been handled outside the Committee on a case-by-case basis. Both the management structure and its effectiveness vary from program to program.

COSTS OF ATMOSPHERIC SCIENCES RESEARCH

Total federal expenditures for research in the United States in fiscal year 1976 amounted to about $21.5 billion. Of this amount, about $290 million, or 1.3 percent, was used for research in the atmospheric sciences, according to the Interdepartmental Committee on the Atmospheric Sciences.*

Meteorological research accounted for 60 percent of the $290 million, meteorological satellite development for 14 percent, high atmospheric research for 24 percent, and atmospheric research on the atmosphere of planets other than the earth for 2 percent. The estimated expenditures, according to institutions, are given in Table 4.

Ten federal agencies provide support for atmospheric research. This diversity is a source of over-all strength and stability; however, it means also that the interrelationships of major programs supported by different agencies may not be easily or fully recognized.

UNIVERSITY RESEARCH AND EDUCATION

Table 4 shows that 19 percent of the federal budget in atmospheric research supports university research; another 15 percent supports

*Interdepartmental Committee for Atmospheric Sciences (1976). *National Atmospheric Sciences Program: Fiscal Year 1977* (ICAS 20-FY77). Federal Council for Science and Technology, Executive Office of the President, Washington, D.C.

TABLE 4 Atmospheric Sciences Expenditures by Type of Institution, Fiscal Year 1976[a]

Type of Institution	$ Millions
Government	$112
Industry	79
Universities	56
Nonprofit Research Institutions	43
	$290

[a]From Interdepartmental Committee for Atmospheric Sciences (1976). *National Atmospheric Sciences Program: Fiscal Year 1977* (ICAS 20-FY77). Federal Council for Science and Technology, Executive Office of the President, Washington, D.C.

research in nonprofit research institutions. In interpreting these quantities, it should be noted that universities and research institutions employ more than half of the Ph.D.-level scientists and contribute more than half of the publications to the scientific literature. These comparisons do not show that government research is less efficient than university research—the budgets of certain government agencies are heavily weighted by the costs of satellites, computers, and large, multiengine aircraft—but they do show that in order to mount effective research programs that place heavy demands on scientific manpower and facilities, the research efforts of federal agencies and nongovernment institutions must be coupled. This is by no means a simple matter, and each major program may well require a different sort of organizational mechanism.

Federal budget support for graduate education and university research in the atmospheric sciences had increased from less than $3 million in 1958 to approximately $42 million in 1967, when the rate of increase stopped rather abruptly. During this same period, there was a greater than threefold increase in the number of university departments offering the Ph.D. degree, and a comparable increase in the number of Ph.D.'s granted per year. It also had the effect of expanding the disciplinary boundaries of the subject and of attracting into the field scientists and students from neighboring disciplines. This was largely the result of a deliberate policy. Scientific manpower in the atmospheric sciences was perceived in 1958 as too limited to meet its challenges and as isolated from the supporting disciplines of physics, chemistry, engineering, and mathematics. In responding to recommendations of a 1958 report by the National Academy of Sciences, the policy of the National Science Foundation was to encourage atmospheric science research of good quality in departments of engineering or physics and to stimulate the development of additional departments and new doctoral degree programs within existing departments. The policy of deliberate expansion of the boundaries of the field has resulted in bringing many scientists into the atmospheric sciences from physics, engineering, and other related fields. The national research program could not have grown as it has in the past 15 years without the contributions of these scientific immigrants. They have played major roles in developing sophisticated instrumentation, especially meteorological satellites and remote-sensing equipment, in mathematical modeling of atmospheric phenomena, and in research management.

Recent reviews of manpower in the atmospheric sciences indicate that degrees granted and jobs available are in rough balance. Looking ahead, one recognizes great uncertainties; and solution of the problems of food, energy, and natural hazard will surely require larger numbers of well-trained atmospheric scientists. The central manpower problems, however, are not so much concerned with numbers as with quality of training and distribution of specialties. In this connection, it may be appropriate at this time to review the policy of deliberate expansion of the field that has contributed so much to present research capabilities.

The universities' research capabilities are substantially extended through the programs of the National Center for Atmospheric Research (NCAR). This center is managed by the University Corporation for Atmospheric Research, a consortium of universities (now 45 in number) having Ph.D. programs in the atmospheric sciences. The central function of NCAR is to organize and carry out, jointly with university faculty and students, research on major atmospheric problems too large to be appropriate to universities. The Center operates a large computer, which is linked to universities in many parts of the country, a fleet of instrumented aircraft, weather radars, and other instruments. In addition to atmospheric research, NCAR supports the High Altitude Observatory, which is devoted to solar physics research, and the National Scientific Balloon Facility which provides high-altitude balloon capability to astronomers, cosmic-ray physicists, and solar physicists, as well as to atmospheric scientists.

The National Science Foundation has been the dominant factor in the dramatic growth and vitalization of research capability that has occurred in the atmospheric sciences. Mere tabulations of agency budgets do not adequately reflect the creative role played by the

Foundation's encouragement and support for quality research at universities and at NCAR over the past 15 years. Scientists recently trained through National Science Foundation research support are adding strength to agency research and operational programs throughout the government and within industry, as well as at universities and research institutions. This new and growing strength is critical to the nation's ability to deal effectively with the complex problems of the atmospheric sciences.

MANAGEMENT AND COORDINATION

In a formal sense, management and coordination at the policy level have been the prerogatives of the Federal Council for Science and Technology (now the Federal Coordinating Council for Science, Engineering, and Technology) and, in the atmospheric sciences, of its subordinate, the Interdepartmental Committee for Atmospheric Research. In this section, we will discuss briefly various procedures employed in the management and coordination of five major programs: the Global Atmospheric Research Program (GARP), weather modification, air quality, climate change, and stratospheric ozone depletion. Although these are separate programs, it is important to recognize that they are also strongly interrelated and that understanding gained from one program may be vitally important to another.

GARP is a major program for which the crucial policy decisions have already been made; consequently, it can be examined as an example of how the system has worked. Weather modification is an example of another sort. Its history is as old as GARP's, but the program is in disarray as a result of inadequate or inappropriate decisions. Research programs in air quality, climate change, and stratospheric ozone depletion are at relatively immature stages, and most of the important policy decisions probably lie ahead. They may be in a position to benefit from the lessons of the past.

The Global Atmospheric Research Program

This international program has two major purposes, extending the range and improving the accuracy of weather forecasts and understanding the physical basis of climate. It consists of a series of rather complex subprograms requiring coordinated planning, execution, and data analysis at both the international and national level. The U.S. budgetary support was approximately $23 million in fiscal year 1976. Indirect support for the related World Weather Watch and the development of technological systems amount to an additional $195 million.

Development of the GARP has moved rather systematically through a series of vital stages: (1) mathematical models of the general circulation of the atmosphere had shown by 1956 that the theory was outstripping observation and that understanding of the global circulation was data-limited, rather than theory-limited; (2) weather satellites demonstrated the capability for making global observations; (3) the scientific concept of the experiment, as proposed by Jule G. Charney of Massachusetts Institute of Technology, received high-level national and international endorsements; (4) a feasibility study and then detailed research plans were drafted under the auspices of the National Research Council; and (5) support of the Congress and the federal agencies was secured, and the necessary funding was obtained. While these steps were being taken, a series of vital technological developments continued, and the first several of the planned sequence of field experiments have been carried out. The first global experiment is scheduled for 1978–1979.

The National Oceanic and Atmospheric Administration has served as lead agency for GARP and the related world weather programs. Thus, the agency most concerned with atmospheric observation and weather-prediction services has a major responsibility for managing and funding the GARP and its subprograms. Coordination among agencies is carried out under the Interagency Committee for the World Weather Program, and coordination of operational matters is the responsibility of the Office of the Federal Coordinator for Meteorological Services and Supporting Research, which reports to the Secretary of Commerce. All of these arrangements were made in the mid-1960's, when science policy decisions were shaped and guided under the auspices of the Office of Science and Technology and the President's Science Advisory Committee.

GARP research, carried out by university and research institute scientists, is supported by the National Science Foundation, through a specific line item in its budget. In 1964, President Johnson specifically endorsed the World Weather Program in a speech at Holy Cross University; and, in 1968, the Congress, through Concurrent Senate Resolution 67, endorsed the Program and required the President to report to the Congress annually on progress. Congressional endorsement helps to keep the program visible to agency administrators, the Congress, and the President, and goes far to ensure continuing support. The scientific community, through two committees of the National Research Council, has been closely involved in each stage of the GARP from early planning through the individual subprograms to evaluation of results.

GARP planning has not always proceeded as directly and smoothly as this summary suggests, and the sequence of advances have not always appeared as clearly marked as they are in retrospect. A more detailed review would include a number of specific difficulties and differences of view that could have derailed the Program if they had not been handled wisely. The key here appears to have been broad participation of nongovernmental agencies and government scientists and careful preparation, both scientifically and administratively.

Weather Modification

Weather modification has been a controversial subject among scientists, government administrators, commercial operators, and the general public for more than 30 years. Although no coherent national program in weather modification exists, federal agencies supported research, amounting to about $15 million in fiscal year 1975, down from $20 million in fiscal year 1973. Operational programs, carried out by commercial operators in the United States and overseas, amounted to about $3.5 million in 1972. In October 1976, the National Weather Modification Policy Act of 1976 (Public Law 94-859) was enacted, requiring a study of weather modification to be carried out under the Secretary of Commerce.

The first comprehensive analysis of cloud-seeding data concluded in 1957 that, in some circumstances, orographic winter precipitation could be increased by about 10 to 15 percent through seeding. After 20 years this conclusion still appears to have validity, but snowfall augmentation over mountains still cannot be regarded as a confirmed, reliable technology. Although there is evidence for optimism, economically significant increases in precipitation from cumulus clouds and from cyclonic storms over flat terrain still are unproven, reduction of winds in hurricanes has not yet been convincingly demonstrated, and hail suppression remains a research enterprise, rather than a confirmed technology in the United States. It is necessary to ask what has constrained advances in weather modification, and what must be done to bring about a more effective program to investigate and develop procedures for beneficially changing the weather.

Research on the physics of clouds is inherently difficult because of the complexities of cloud nucleation and microphysics, because of difficulties of observation, and because of the natural variability of clouds and cloud systems. There have been impressive advances in the understanding of clouds and precipitation, but much more could have been done if there were more funds, facilities, and scientific competence. Progress also has been constrained over the past 25 years by inadequate planning, coordination, and management.

Despite frequent references to the contrary, a national program in weather modification does not exist. Neither the legislative nor the executive branch of the government treats the development of capabilities to increase water supplies by seeding or to modify hurricanes as national policy issues, even though successful efforts clearly would have enormous impact on national policies relating to food, energy, and natural hazards. There is no lead agency for weather modification, although reports of the Committee on Atmospheric Sciences of the National Research Council, the General Accounting Office, the National Advisory Committee on Oceans and Atmosphere, and others have identified this as a critical need. As a natural consequence, budget decisions are responsive, at best, to individual agency perceptions of the national interest. In early 1973, as a result of actions by the Office of Management and Budget, five of seven weather-modification programs, which had been identified by the Federal Council as national projects, were terminated or suspended without scientific review.

A second factor complicating the field of weather modification is the existence of statutes adopted by about 30 states, which are inconsistent and contradictory. This has led to programs that are uncoordinated, so that it is possible for field programs in one state to influence neighboring states and to contaminate experimental results.

A third important consideration is the presence of potentially serious international issues that call for a coherent national policy. For example, in order to conduct hurricane-modification experiments anywhere, it is considered appropriate to enter into agreements with those countries likely to be affected. Also, military use of weather modification stimulates international suspicions and could pose serious problems for important scientific programs, as well as for other aspects of foreign policy.

Finally, U.S. commercial cloud seeders are operating in foreign countries, in many cases with minimal facilities, monitoring, and supervision.

The following institutional changes appear as essential if these deficiencies are to be overcome: (a) a single agency with broad competence and responsibilities in the atmospheric sciences should be designated as the lead agency for interagency programs in weather modification and should be provided with adequate, stable research support; and (b) federal legislation should establish a basis for resolving interstate and international issues relating to weather modification.

Air Quality

The deterioration of atmospheric quality, resulting from human activities, and its deleterious effects have received a great deal of attention over the last decade. As industrialization and world population continue to increase, we must anticipate that problems associated with atmospheric quality will take on greater urgency. In some instances we know enough to justify concern and rational response; in many others, we know only that we should be alert. For example, emission of specific toxic materials by industry poses direct threats to health in areas close to the sources. Emission of sulfur from highly industrialized regions results in markedly acidic rainfall in downwind regions. The consequences may include widespread damage to forests, agriculture, and aquatic fauna and flora. Deleterious effects of sulfur on human health are under investigation. As noted in earlier chapters, increased emissions into the atmosphere, particularly of carbon dioxide, pose long-range threats to the global climate.

Responsibility for monitoring and research in air quality or air pollution for many years has been divided among the National Oceanic and Atmospheric Administration (NOAA), the Environmental Protection Agency (EPA), and the Energy Research and Development Administration (ERDA). The missions of the last two agencies include understanding how air quality is changing, or might change, and what should be done to avoid undesirable alterations. NOAA is involved for several reasons: (a) the ability of the atmosphere to transport, diffuse, and deposit pollutants depends on physical and chemical properties of the atmosphere, which is NOAA's responsibility to observe and to understand; (b) air-quality monitoring can be carried on efficiently in many cases at weather stations; and (c) pollutants may in some cases modify weather processes.

An outstanding example of a successful air-quality research program has been the Metropolitan Meteorological Experiment, a five-year program of field experiments addressed to the effects of urban pollutants and energy on clouds and precipitation in the area of St. Louis. The experiment was planned in 1970 by principal investigators from the University of Chicago, the Illinois State Water Survey, Argonne National Laboratory, and the University of Wyoming. It has been supported largely by various government agencies at a level amounting to about $1 million per year. The program is composed of components carried out essentially independently under the several principal investigators within a common schedule and with voluntary coordination. This mode of management has been highly effective in this program, but there could be severe difficulties in applying it to programs requiring higher degrees of interdependence.

When the EPA was created on December 2, 1970, it was widely assumed that it would assume leadership in air-quality research and monitoring. Ambitious plans were announced for a grant program to develop university centers specializing in air-quality research. Although a few grants were made, the original plans have not been carried out. A major research program, called the Regional Air Pollution Study, was planned more than five years ago as an interagency program to be carried out in the St. Louis area and to be coordinated with the Metropolitan Meteorological Experiment. The EPA had the lead-agency responsibility, but the program was plagued with a series of delays and administrative difficulties, and the original interagency concept has been abandoned. There are several lessons to be learned from this experience, although a more thorough analysis would be necessary to draw firm conclusions. When regulatory and research responsibilities are combined within a single agency, it is difficult to do both jobs competently; broad participation of other agencies and the nongovernmental scientific community would be of major assistance in planning and managing major atmospheric research programs.

Stratospheric Ozone Depletion

In 1971, the Congress rejected the Administration's effort to build prototypes of the Boeing SST, thus terminating, at least for the time being, development of a supersonic transport plane in the United States. Among the issues that were cited by opponents as reasons for not going ahead with the SST were (a) possible effects of effluents on climate and (b) possible depletion of stratospheric ozone and the consequent increase in the incidence of skin cancer. The information available was inadequate, either to counter these arguments or to verify them. How important these issues were in determining the outcome of congressional action is hard to say, but they surely were perceived by the Congress and the Department of Transportation as potentially important issues that had to be resolved before a second attempt at securing approval for an SST development program.

Even before rejection of the SST by Congress, the Department of Transportation, in response to the urging of Senator Henry Jackson, had drawn up plans for a Climatic Impact Assessment Program. The program was launched in 1971 with a rigid schedule that specified completion on December 31, 1974. Planning and management of the program was carried out within a unit attached to the Office of the Assistant Secretary for Systems Development and Technology. The program apparently was viewed as a narrowly defined engineering task with goals set in the form of reports due on specific dates. A committee of the National Research Council was responsible for reviewing the program, and it emphasized the need to regard the program as a scientific, rather than only an engineering program. The NAS report,* representing an independent assessment, helped greatly to specify the nature of the stratospheric problem and to identify areas requiring further efforts.

Completion of the Climatic Impact Assessment Program coincided with recognition that stratospheric ozone may be depleted by chlorofluorocarbons and other commercially produced stable compounds that are ultimately released into the atmosphere. As a result of the program, the potential danger of chlorofluorocarbons and other gases was probably recognized earlier than it otherwise would have been, and the nation is better prepared to set about getting definitive information on how to deal with this new threat.

The Climatic Impact Assessment Program was criticized by certain members of the scientific community for not giving adequate attention to the need for plans to continue promising research at the end of the original program. This point was proven valid. Depletion of stratospheric ozone depends on cumulative processes that will require five years or more for directly observable effects. As a result of the new threat to the ozone layer posed by chlorofluorocarbons and similar substances, there is no question but that research should continue with substantial support. The National Aeronautics and Space Administration (NASA) has undertaken an expanded role in upper-atmosphere research and has become the lead agency for a program to develop new approaches to instrumentation and measuring systems necessary to assess the effect of chlorofluoromethanes and other compounds in the atmosphere. In this way, NASA has become the key agency in pursuit of the ozone problem.

Although the problems of stratospheric modification have so far been treated exclusively on the national level, they obviously must be

*Climatic Impact Committee (1975). *Environmental Impact of Stratospheric Flight: Biologic and Climatic Effects of Aircraft Emissions in the Stratosphere*. National Academy of Sciences, Washington, D.C.

approached on the international level as well. The potential dangers of ozone depletion apply to all nations, and joint action will be essential to provide comprehensive protection. In this connection, a potential military threat may also exist, and must be better understood, and soon. A recent NAS study* has pointed out that explosion of half the world's nuclear stockpile could reduce stratospheric ozone by 30 to 70 percent in the northern hemisphere and that such an ozone loss could have serious adverse effects on crops and natural ecosystems, as well as on humans. Suggestions have been made that increased production and use of nitrogen fertilizer may result in depletion of stratospheric ozone. This problem obviously demands attention at the international level, as well as the national level.

Climate Change

A great upsurge of interest in climate is occurring simultaneously in the scientific community and in policy levels of government. An early focal point of this new interest in government in 1972 and 1973 was the National Security Council reflecting concern with strategic military aspects. Responsibility has been shifted to the Domestic Council, and an interagency subcommittee has been created. However, neither the Security Council nor the Domestic Council is staffed to deal adequately with scientific issues, and neither has a background of experience in collecting and assessing scientific advice on climate or atmospheric research. The moves that have occurred in the governmental councils have been *ad hoc* efforts to deal with a critical issue in the absence of a science policy component in the Executive Office of the President. Interest of the Congress in climate is illustrated by the hearings held in May 1976 by the Committee on Science and Technology on H.R. 13736, a bill authorizing establishment of a national climate program.

In 1971, the Committee on Atmospheric Sciences of the National Research Council urged special research efforts to increase understanding of climate, and in 1975 the United States Committee for the Global Atmospheric Research Program issued a report, entitled *Understanding Climatic Change: A Program for Action*. Early versions of this report, as well as others, served as major sources of information for the interagency committee program that has been presented to the Domestic Council. In December 1974, the Domestic Council issued a

*NRC Committee to Study Long-Term Worldwide Effects of Multiple Nuclear-Weapons Detonations (1975). *Long-Term Worldwide Effects of Multiple Nuclear-Weapons Detonations*. National Academy of Sciences, Washington, D.C.

report, entitled *A United States Climate Program*, calling for the establishment of a national program of climate research. As of 1976, such a program still has not come into being.

Although there is strong advocacy for expanding climate research and hopes for greater funding, there are reasons for serious concern. First, climate change is a complex field in which hypotheses are easier to formulate than to verify. For this reason, it may be difficult to maintain the balanced perspective that is essential to long-term research. Second, responsibility for surveillance of the field is shared in an obscure manner between the Domestic Council and the Federal Coordinating Council for Science, Engineering, and Technology. Third, there is no designated lead agency responsible for coordinating the various agency programs of climate research and of exerting leadership within government.

Summary

The atmospheric sciences are fundamentally healthy and strong. Impressive scientific strides have been made in the past 15 years, and important contributions to the nation's efforts to cope with the problems of food and energy production and natural hazard are in prospect. However, serious deficiencies exist in coordination and management of major programs, which much be rectified if the full potential contribution of the atmospheric sciences is to be realized.

The five major research programs reviewed in this report have demonstrated some of the strengths and some of the weaknesses present in major atmospheric research programs. These five programs are complex and administratively distinct from each other, but the following generalizations may be useful guides to improvements in management and coordination of these and other major research programs.

1. The atmospheric programs discussed here require collaboration of several federal agencies and the participation of the nongovernmental scientific community. This perception appears not to be shared by some agency administrators, who are likely to respond strongly to issues affecting agency strength, survival, and competition.
2. The influence of the scientific community in planning and in contributing to major research programs has been effective only when the lead agency has recognized a need for advice.
3. The new Office of Science and Technology Policy, created by the National Science and Technology Policy, Organization and Priorities

Act of 1976, which restores science capability to the Executive Office of the President, can play a vital role by promoting interagency cooperation; identifying and designating appropriate lead agencies; and ensuring the integrity of programs whose objectives fall outside the separate agency missions or that overlap the responsibilities of more than one agency.

Concluding Statement

Atmospheric research contributes to the missions of many government agencies, and programs of research are interwoven through the various agencies in different ways. Therefore, the recommendations of this chapter, which imply institutional changes in government and in the relation of the scientific community to government, may impinge on the agencies in different ways. Also, the problems of research management discussed in this chapter may have their counterparts in other environmental fields. For these reasons, ideal management structures will be especially difficult to design and to achieve. Yet, significant and critical advances along the scientific front are likely to occur only if appropriate institutional changes can be made.

Clearly, a necessary next step should be to direct broad attention at the nation's capacity to meet complex environmental problems. The division of responsibilities among federal agencies, their research capabilities, and their leadership and coordination of interagency programs should be reviewed. The allocation of resources to agency in-house research and to contract and grant research should be examined. And critical assessments of the role of the nongovernmental scientific community in programs of national importance should be made. The results of such a broad review should contribute to research programs in the atmospheric sciences and in other fields that are well planned, coordinated, and managed at the same time that they help to assure the public that major societal problems are being attacked as effectively as possible by the scientific community and the federal government.

Additional Reading

American Meteorological Society (1973). Policy Statement of the AMS on Weather Forecasting, *Bull. Am. Meteorol. Soc. 54*(1):47–48.

Battan, L. J. (1961). *The Nature of Violent Storms*. Science Study Series. Doubleday & Co., Inc., Garden City, N.Y., 158 pp.

Battan, L. J. (1969). *Harvesting the Clouds*. Doubleday & Co., Inc., Garden City, N.Y., 148 pp.

Budyko, M. I. (1974). *Climate and Life*. English Edition. D. H. Miller, ed., International Geophysics Series, Vol. 18. Academic Press, New York, 508 pp.

Committee on Atmospheric Sciences (1973). *Weather and Climate Modification: Problems and Progress*. National Academy of Sciences, Washington, D.C., 258 pp.

Committee on Impacts of Stratospheric Change (1976). *Halocarbons: Environmental Effects of Chlorofluoromethane Release*. National Academy of Sciences, Washington, D.C., 125 pp.

Fleagle, R. G. (1969). *Weather Modification: Science and Public Policy*. University of Washington Press, Seattle, Washington, and London, England, 147 pp.

Hess, W. H., ed. (1974). *Weather and Climate Modification*. John Wiley & Sons, New York, 842 pp.

Kellogg, W. W., R. D. Cadle, E. R. Allen, A. L. Lazrus, and E. A. Martell (1972). The Sulfur Cycle, *Science 175*:587–597.

Machta, L., and E. Hughes (1970). Atmospheric Oxygen in 1967 to 1970, *Science 168*:1582–1584.

Matthews, W. H., W. W. Kellogg, and G. D. Robinson, eds. (1971). *Inadvertent Climate Modification, Study of Man's Impact on Climate* (SMIC). MIT Press, Cambridge, Mass. 308 pp.

Mathews, S. W. (1976). What's Happening to Our Climate? *National Geographic 150*(5):576–615.

Mintz, Y. (1970). The Four Basic Requirements for Weather Prediction. *Global Weather Prediction: The Coming Revolution.* Lusignan & Kiely, eds. Holt, Rinehart & Winston, New York, 307 pp.

Miyakoda, K. (1974). Numerical Weather Prediction. *Am. Scientist 62*(5):564–574.

Taubenfeld, H. J. (1970). *Controlling the Weather: A Study of Law and Regulatory Processes.* The Dunellen Publishing Co., Inc., New York, 275 pp.

Thompson, J. C. (1972). *The Potential Economic Benefits of Improvements in Weather Forecasting.* Final Report, Dept. of Meteorol., California State U., San Jose, Calif., 80 pp.

Thompson, P. D., and R. Obrien (1968). *Weather.* Life Science Library, Time-Life Books, New York, 200 pp.

U.S. Committee for the Global Atmospheric Research Program (1975). *Understanding Climatic Change: A Program for Action.* National Academy of Sciences, Washington, D.C., 239 pp.

White, G. F., and J. E. Haas (1975). *Assessment of Research on Natural Hazards.* MIT Press, Cambridge, Mass., and London, England, 487 pp.

Wilson, C. L., and W. H. Matthews, eds. (1970). *Man's Impact on the Global Environment.* Study of Critical Environmental Problems (SCEP). MIT Press, Cambridge, Mass. 319 pp.